MATHS NOW!

intermediate

GCSE

1

MATHS NOW!
National Writing Group

JOHN MURRAY

Acknowledgements

The authors and publisher would like to thank all the teachers, schools and advisers who evaluated *Maths Now!* and whose comments contributed so much to this final version.

Particular thanks go to: Philip Chaffé, King Edward VI School, Lichfield; John D Collins, Education Consultant and Inspector of Schools; Patrick Gallagher, Convent of Jesus and Mary RC High School, London; Ian Gregory, Blakeston Community School, Stockton-on-Tees; David J McLaren, Consultant in Mathematical Education; David Bullock, Frodsham High School, Warrington; Peter Marks, Ilfracombe College, Devon; Kevin Pankhurst, Pilton Community College, Barnstaple; Kim O'Driscoll-Tole, University of Strathclyde; Mrs T Stephens, Bedwas School, Gwent.

Photo acknowlegements

cover Images Colour Library; **p.24** *both* © John Townson/Creation; **p.25** *both* Ancient Art and Architecture Collection Ltd; **p.50** *t* Proportions of the human figure, c.1492 (Vitruvian Man) (pen and ink on paper) by Leonardo da Vinci (1452–1519) Galleria dell'Accademia, Venice, Italy/Bridgeman Art Library; *b* Portrait of a Bearded Man, possibly a Self Portrait, c.1513 (red chalk on paper) by Leonardo da Vinci (1452–1519) Biblioteca Reale, Turin, Italy/Bridgeman Art Library; **p.51** *all* Science Museum/Science and Society Picture Library; **p.55** *tl* Science Museum/Science and Society Picture Library; *tr* Professor Peter Goddard/Science Photo Library; *bl* Rex Features Limited; **p.66** *both* © John Townson/Creation; **p.68** Rex Features Limited; **p.72** © John Townson/Creation; **p.77** *l* © Macduff Everton/Corbis; *r* © Jim Holmes/Axiom; **p.88** Hulton Archives; **p.102** *all* © John Townson/Creation; **p.113** *both* © Chris Caldicott/Axiom; **p.134** *both* Werner Forman Archive; **p.155** *tl* NASA/Science Photo Library; *tr* © John Townson/Creation; *bl* David Parker/Science Photo Library; *br* English Heritage Photographic Library, © Skyscan Balloon Photography; **p.161** Hulton Archives; **p.179** © National Maritime Museum Picture Library; **p.200** © John Townson/Creation; **p.214** *tl* © Donald Cooper/Photostage; *tr* David A. Hardy/Science Photo Library; *bl* Rex Features Limited; *br* Rex Features Limited.

t = top, *b* = bottom, *l* = left, *r* = right

© R. Pimentel and T. Wall/MATHS NOW! National Writing Group 2001

First published in 2001
by John Murray (Publishers) Ltd
50 Albemarle Street
London W1S 4BD

Layouts by Stephen Rowling/springworks
Illustrations by Oxford Illustrators Ltd
Cover design by John Townson/Creation

Typeset in 10/12pt Times by Wearset, Boldon, Tyne and Wear
Printed and bound in Spain by Book Print S.L., Barcelona

A CIP catalogue record for this book is available from the British Library

ISBN 0 7195 7447 1
Maths Now! GCSE Intermediate 1 Teacher's Resource Book 0 7195 7448 X

MATHS NOW!

GCSE

intermediate

1

Contents

Introduction

In this year you will be at the start of your GCSE course. This book, along with Intermediate 2, contains all the material to prepare you for your examinations.

Mathematics, however, is not just about passing exams. Therefore this book also touches on some of the historical roots of mathematics. You will also find problems that require the use of ICT and also some longer, investigative style questions.

We hope that you will approach the work in this book in a positive way, expecting both difficulty and enjoyment. We hope that you gain a sense of achievement and that this achievement is reflected in the highest grade in your GCSE examination of which you are capable.

Symbols

The symbols used in the Student's Book are as follows:

 Use a calculator

 Do not use a calculator

 Ma1

1 Integers, factors and primes

Our present number system has a long history, originating from the Indian Brahmi numerals around 300 BC.

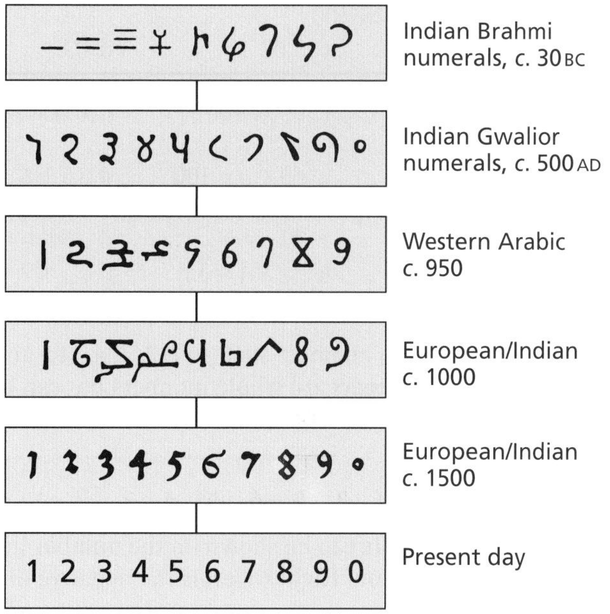
Indian Brahmi numerals, c. 30 BC
Indian Gwalior numerals, c. 500 AD
Western Arabic c. 950
European/Indian c. 1000
European/Indian c. 1500
1 2 3 4 5 6 7 8 9 0
Present day

Revision

- Multiplying by 10, 100, 1000 results in the digits moving 1, 2 or 3 places to the left respectively.

  $28 \times 10 = 280$ $\quad$ $34.56 \times 10 = 345.6$
  $28 \times 100 = 2800$ $\quad$ $34.56 \times 100 = 3456$
  $28 \times 1000 = 28\,000$ $\quad$ $34.56 \times 1000 = 34\,560$

- Similarly, dividing by 10, 100, 1000 results in the digits moving 1, 2 or 3 places to the right respectively.

  $28 \div 10 = 2.8$ $\quad$ $34.56 \div 10 = 3.456$
  $28 \div 100 = 0.28$ $\quad$ $34.56 \div 100 = 0.3456$
  $28 \div 1000 = 0.028$ $\quad$ $34.56 \div 1000 = 0.03456$

- $27 + 73 = 100$.
  So 73 is said to be the **complement** of 27 to 100.

Exercise 1.1

1 Multiply the following numbers by 10.
a 63 **b** 4.6 **c** 0.84 **d** 0.065 **e** 1.07

2 Multiply the following numbers by 100.
a 45 **b** 7.2 **c** 0.96 **d** 0.0485 **e** 6.033

3 Multiply the following numbers without using a calculator.
  a 46×10000 **b** 6.8×1000 **c** 3.8×100000
  d 0.0084×10000 **e** 0.7×100000

4 Divide the following numbers by 10.
  a 680 **b** 72 **c** 8.9 **d** 0.64 **e** 0.054

5 Divide the following numbers by 100.
  a 3500 **b** 655 **c** 5.62 **d** 0.8 **e** 0.034

6 Find the value of:
  a $6.4 \div 10000$ **b** $46 \div 1000$ **c** $950 \div 100$ **d** $0.0845 \div 10000$ **e** $4 \div 1000$

7 Find the value of each letter.
  a $43 + a = 100$ **b** $51 + b = 100$ **c** $69 + c = 100$ **d** $16 + d = 100$ **e** $e + 89 = 100$

8 Copy and complete the following.
  a $7 \times 8 = \underline{}$ **b** $6 \times 9 = \underline{}$ **c** $72 \div 9 = \underline{}$ **d** $36 \div 4 = \underline{}$ **e** $8 \times \underline{} = 64$
  f $7 \times \underline{} = 42$ **g** $5 \times \underline{} = 45$ **h** $\underline{} \times 4 = 32$ **i** $\underline{} \times 9 = 63$ **j** $\underline{} \div 6 = 6$

Whole numbers such as 1, 2, 3, 4, etc. are called **natural numbers**.

Integers are whole numbers but can be positive, negative or zero. Examples of integers are 7, 12, −3 and −5.

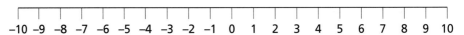

Integers can be shown on the number line above. Left to right is the positive direction. Right to left is the negative direction.

Adding integers

Examples

Use the number line above to add $(+4)$ and (-2).

Start at $(+4)$ and move 2 in the negative direction.

$$(+4) + (-2) = (+2) \quad \text{or } 2$$

> **Remember:**
> *For **addition**, start from one number. Follow with the other number.*

Use the number line to add (-5) and $(+7)$.

Start at (-5) and move 7 in the positive direction.

$$(-5) + (+7) = (+2) \quad \text{or } 2$$

Use the number line to add (-1) and (-3).

Start at -1 and move 3 in the negative direction.

$$(-1) + (-3) = (-4) \quad \text{or } -4$$

Use the number line to add $(+2)$, (-3) and $(+5)$.

Start at $+2$ and move 3 in the negative direction, then 5 in the positive direction.

$$(+2) + (-3) + (+5) = (+4) \quad \text{or } 4$$

Use the number line to add (-2), $(+3)$ and (-4).

Start at -2 and move 3 in the positive direction, then 4 in the negative direction.

$$(-2) + (+3) + (-4) = (-3) \quad \text{or } -3$$

Exercise 1.2

Draw a number line from -8 to $+8$. Use it to answer the questions below.

1 a $(+5) + (-2)$ **b** $(+3) + (-3)$ **c** $(+8) + (-5)$
2 a $(-5) + (+3)$ **b** $(-8) + (+7)$ **c** $(-2) + (-6)$
3 a $(-3) + (-3)$ **b** $(-3) + (-2)$ **c** $(-2) + (-6)$
4 a $(-3) + (+3) + (-1)$ **b** $(-4) + (-4) + (+6)$ **c** $(+1) + (-4) + (-3)$

Subtracting integers

Examples

Use the number line above to calculate $(+6) - (+2)$.

Start at $(+2)$. Then $(+6)$ is 4 in the positive direction. So

$$(+6) - (+2) = +4$$

Remember:
*For **subtraction**, always start at the number being subtracted. Find the direction and distance to the other number.*

Use the number line to calculate $(-2) - (+3)$.

Start at $(+3)$. Then (-2) is 5 in the negative direction. So

$$(-2) - (+3) = -5$$

Use the number line to calculate $(-3) - (-6)$.

Start at (-6). Then (-3) is 3 in the positive direction. So

$$(-3) - (-6) = +3$$

Exercise 1.3

Using a number line if necessary, work out the answer to each of the following calculations.

1 a $(+4) - (+3)$ **b** $(+5) - (+2)$ **c** $(+7) - (+5)$
   d $(+4) - (+1)$ **e** $(+8) - (+2)$ **f** $(+6) - (+3)$
2 a $(+2) - (+3)$ **b** $(+4) - (+5)$ **c** $(+3) - (+6)$
   d $(+2) - (+7)$ **e** $(+4) - (+5)$ **f** $(+1) - (+2)$
3 a $(-2) - (+4)$ **b** $(-3) - (+5)$ **c** $(-6) - (+8)$
   d $(-1) - (+6)$ **e** $(-3) - (+3)$ **f** $(-5) - (+7)$
4 a $(-8) - (-4)$ **b** $(-2) - (-1)$ **c** $(-5) - (-2)$
   d $(-7) - (-6)$ **e** $(-6) - (-3)$ **f** $(-7) - (-5)$
5 a $(-2) - (-3) + (-4)$ **b** $(-1) - (-4) - (+1)$ **c** $(-3) - (-7) + (-7)$
   d $(-5) - (-6) - (-3)$ **e** $(-4) - (-8) - (-5)$ **f** $(-1) - (-3) + (+2)$

Multiplication and division of integers

For multiplication of integers, we use the idea of repeated addition.

$(+3) \times (+2)$ means 3 lots of $+2$ or $(+2) + (+2) + (+2) = (+6)$
$(+3) \times (-2)$ means 3 lots of -2 or $(-2) + (-2) + (-2) = (-6)$
$(-3) \times (+2)$ means -3 lots of $+2$ or $-((+2) + (+2) + (+2)) = -(+6) = (-6)$
$(-3) \times (-2)$ means -3 lots of -2 or $-((-2) + (-2) + (-2)) = -(-6) = (+6)$

Exercise 1.4

1 Calculate the answer to each of the following.
   a $(+9) \times (-5)$ **b** $(+8) \times (-5)$ **c** $(+7) \times (-6)$

2 Calculate the answer to each of the following.
   a $(-5) \times (+3)$ **b** $(-6) \times (+9)$ **c** $(-8) \times (+8)$

3 Copy and complete the multiplication table below.

$\times$	-3	-2	-1	0	$+1$	$+2$	$+3$
+3		-6					$+9$
+2							
+1					$+1$		
0	0			0			
−1					-2		
−2							
−3							

4 Calculate the answer to each of the following.
   a $(+3) \times (+2)$ **b** $(-2) \times (+1)$ **c** $(+3) \times (-2)$
   d $(-2) \times (-3)$ **e** $(-3) \times (-1)$ **f** $(-2) \times (-2)$

Remember:
When multiplying:

- *a positive number by a positive number, the result is positive;*
- *a negative number by a positive number, the result is negative;*
- *a positive number by a negative number, the result is negative;*
- *a negative number by a negative number, the result is positive.*

> Like signs give +, unlike signs give −.

The rules for division of integers are the same as those above for multiplication. When both quantities are the same sign, the result is positive; when one is positive and the other is negative, the result is negative.

$$(+12) \div (-3) = -4 \qquad (-12) \div (-3) = 4$$

Exercise 1.5

1 Calculate:

a $(+15) \div (+3)$ **b** $(+15) \div (-3)$ **c** $(-15) \div (+3)$

d $(-15) \div (-3)$ **e** $(-28) \div (-7)$ **f** $(-45) \div (+9)$

2 Find the missing number to make each of the statements below true.

a $(\_) \times (+5) = (+15)$ **b** $(\_) \times (-3) = (-15)$

c $(\_) \times (-4) = (-20)$ **d** $(-7) \times (\_) = (-21)$

e $(+5) \times (\_) = (-40)$ **f** $(-7) \times (\_) = (+21)$

3 The table below gives pairs of numbers x and y which add together to make 10: that is, $x + y = 10$. Copy and complete the table.

x	+5	+4	+3	+2	+1	0	−1	−2	−3	−4	−5
y											

4 If $p + q = -2$, copy and complete the table below.

p	+5	+4	+3	+2	+1	0	−1	−2	−3	−4	−5
q											

5 If $xy = +12$, copy and complete the table below. (Note that xy means x multiplied by y.)

x	+4	+3	+2	+1	−1	−2	−3	−4
y								

Factors and primes

Prime numbers have fascinated mathematicians for centuries.

Eratosthenes (275–194 BC) identified primes with a 'sieve'.

Mersenne, a French monk, thought he had a clever theory about primes, which proved partly correct.

There is a big cash prize for finding new prime numbers (something to occupy you on a wet Sunday afternoon!).

Teams of mathematicians are now searching for primes using the GIMPS internet.

The **factors** of a number are all the whole numbers (positive integers) which divide exactly into that original number. For example, the factors of 12 are all the numbers which divide into 12 exactly. They are 1, 2, 3, 4, 6, 12.

A **prime number** is one that has only two factors, those being 1 and itself. (Note that, by definition, 1 is *not* a prime number.)

Prime factors are those factors of a number which are also prime numbers. Therefore, the prime factors of 12 are 2 and 3.

Finding the prime factors of a number can be done by constructing a factor tree.

Examples   Find the prime factors of 45.

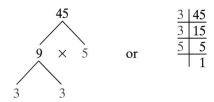

A branch is not continued further when a prime number is reached. So the prime factors of 45 are 3 and 5.

Express 45 as a product of primes.

From the factor tree above we can see that 45 can be written as $3 \times 3 \times 5$ or $3^2 \times 5$.

Exercise 1.6

1 In a 10 by 10 square, write the numbers 1 to 100.
  a Cross out number 1.
  b Cross out all the even numbers after 2 (these have 2 as a factor).
  c Cross out every third number after 3 (these have 3 as a factor).
  d Continue with 5, 7, 11 and 13, then list all the prime numbers less than 100.
2 List all the factors of each of the following numbers and circle the prime factors.
  a 6   **b** 9   **c** 7   **d** 15   **e** 24
  f 36   **g** 35   **h** 25   **i** 42   **j** 100
3 Find the prime factors of each of the following numbers and express the number as a product of prime numbers.
  a 12   **b** 32   **c** 36   **d** 40   **e** 44
  f 56   **g** 16   **h** 39   **i** 231   **j** 63
4 What is meant by the 'sieve of Eratosthenes'?

Highest common factor and lowest common multiple

The factors of 12 are 1, 2, 3, 4, 6 and 12.
The factors of 18 are 1, 2, 3, 6, 9 and 18.

The **highest common factor** (HCF) of 12 and 18 is therefore 6, as it is the largest factor to appear in both groups, i.e. it is common to both.

The multiples of 6 are those numbers in the 6 times table, i.e. 6, 12, 18, 24, 30, etc.
The multiples of 8 are those numbers in the 8 times table, i.e. 8, 16, 24, 32, 40, etc.

The **lowest common multiple** (LCM) of 6 and 8 is therefore 24, as it is the smallest multiple to appear in both groups.

Exercise 1.7

1 Find the highest common factor of each of the following sets of numbers.
- **a** 8, 12
- **b** 10, 25
- **c** 12, 18, 24
- **d** 15, 21, 27
- **e** 36, 63, 108

2 Find the lowest common multiple of each of the following sets of numbers.
- **a** 6, 14
- **b** 4, 15
- **c** 2, 7, 10
- **d** 3, 9, 10
- **e** 3, 7, 11

3 Diagonals ⌒Ma1⌒

How many squares of a grid does a diagonal pass through? For example, on a 4 × 3 grid, the diagonal passes through six squares:

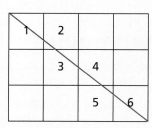

Investigate some more grids and find a relationship between the dimensions of the grid and the number of squares the diagonal passes through.

Rounding

Examples   If 28 617 people attend a gymnastics competition, this figure can be reported to various levels of accuracy.

Round 28 617 to the nearest 10 000.

20 000   30 000
                                                                        28 617

The number line shows 20 000 and 30 000. These are the 'ten thousands' on either side of 28 617. As 28 617 is closer to 30 000, we say that 28 617 is 30 000 rounded to the nearest 10 000.

Round 28 617 to the nearest 1000.

28 000   29 000
                                        28 617

The number line shows 28 000 and 29 000. These are the 'thousands' on either side of 28 617. As 28 617 is closer to 29 000, we say that 28 617 is 29 000 rounded to the nearest 1000.

Remember:
When rounding to the nearest 100, if the digit in the tens column is 5 or more, round up. If it is 4 or less, round down.

Exercise 1.8

1 Round each of the following numbers to the nearest 1000.
   a 68 786   **b** 74 245   **c** 89 000   **d** 4020   **e** 99 500   **f** 999 999
2 Round each of the following numbers to the nearest 100.
   a 78 540   **b** 6858   **c** 14 099   **d** 8084   **e** 950   **f** 2984
3 Round each of the following numbers to the nearest 10.
   a 485   **b** 692   **c** 8847   **d** 83   **e** 4   **f** 997

Significant figures

The use of **significant figures** (s.f.) is another way of 'rounding' an answer. The term 'significant figures' refers to the most important digits in a number.

Example   4367 people attended a basketball match.
   a Give this number correct to one significant figure.
   b Give this number correct to two significant figures.

   a One significant figure implies that only the most important digit is needed in the answer. The number line below shows 4000 and 5000. These are the numbers correct to 1 s.f. either side of 4367.

4000   5000
                                        4367

As 4367 is closer to 4000 we say that 4367 is 4000 correct to 1 s.f.

b Two significant figures implies that only the two most important digits are needed in the answer. The number line below shows 4300 and 4400. These are the numbers correct to 2 s.f. either side of 4367.

4300   4400

4367

As 4367 is closer to 4400 we say that 4367 is 4400 correct to 2 s.f.

Decimal places

A number can also be approximated to a given number of **decimal places** (d.p.). This refers to the number of digits written after the decimal point.

Example | The length of a fossil worm is measured as 7.864 cm. As this is an unnecessary degree of accuracy, give this length correct to one decimal place.

The length needs to be written with one digit after the decimal point. The number line below shows 7.8 and 7.9. These are the numbers correct to 1 d.p. either side of 7.864.

> **Remember:**
> *To round to a certain number of decimal places, look at the digit in the next decimal place along. If it is 5 or more, round up; if it is 4 or less, round down.*

7.8   7.9

7.864

As 7.864 is closer to 7.9 we say that 7.864 is 7.9 correct to 1 d.p.

Exercise 1.9

1 Write each of the following numbers correct to one significant figure.
  a 48 599   **b** 6.38   **c** 0.974   **d** 3620   **e** 9.3 billion
2 Write each of the following numbers correct to two significant figures.
  a 842   **b** 6.92   **c** 0.00156   **d** 0.8104   **e** 6.8 million
3 Give each of the following correct to one decimal place.
  a 5.58   **b** 0.73   **c** 11.86   **d** 157.39   **e** 12.049
4 Give each of the following correct to two decimal places.
  a 6.473   **b** 9.587   **c** 16.476   **d** 0.088   **e** 3.0037

Estimating answers to calculations

Even though many calculations can be done quickly and effectively on a calculator, often an **estimate** for an answer can be a useful check. This is found by rounding each of the numbers so that the calculation becomes relatively straightforward.

Examples Estimate the answer to 57×246.

Below are two possibilities, both of which are acceptable.

$$60 \times 200 = 12\,000$$
$$50 \times 250 = 12\,500$$

Estimate the answer to $6386 \div 27$.

$$6000 \div 30 = 200$$

Estimate the answer to $57 \div 0.48$.

$$60 \div 0.5 = 120$$

Exercise 1.10

1 Without using a calculator, **estimate** the answers to the following.
  a 62×19 **b** 270×12 **c** 55×60
  d 4950×28 **e** 0.8×0.95 **f** 0.184×475

2 Without using a calculator, **estimate** the answers to the following.
  a $3946 \div 18$ **b** $8287 \div 42$ **c** $906 \div 27$
  d $5520 \div 13$ **e** $48 \div 0.12$ **f** $610 \div 0.22$

3 Using estimation, identify which of the following are definitely incorrect. Explain your reasoning clearly.
  a $95 \times 212 = 20\,140$ **b** $44 \times 17 = 748$ **c** $689 \times 413 = 28\,457$

  d $142\,656 \div 8 = 17\,832$ **e** $77.9 \times 22.6 = 2512.54$ **f** $\dfrac{84.2 \times 46}{0.2} = 19\,366$

4 Estimate the areas of the following shapes. Do *not* work out an exact answer.

  a

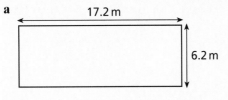

17.2 m   6.2 m

  b

9.7 m   2.6 m   3.1 m   4.8 m

5 Estimate the volume of each of the solids below. Do *not* work out an exact answer.

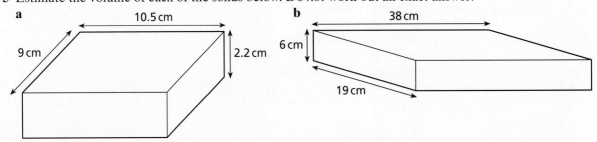

  a   10.5 cm   9 cm   2.2 cm

  b   38 cm   6 cm   19 cm

SUMMARY

By the time you have completed this chapter you should know:

- how to add, subtract, multiply and divide integers
- how to multiply and divide by any power of ten, for example

$$41.5 \div 10 = 4.15 \qquad 41.5 \times 10 = 415$$

- what is meant by **prime numbers**, **prime factors**, the **highest common factor** and the **lowest common multiple**
- how to round numbers to the nearest power of ten or to a given number of **decimal places** and **significant figures**, for example

$$837.6 \text{ is } 840 \text{ to } 2\,\text{s.f.} \qquad 8.376 \text{ is } 8.38 \text{ to } 2\,\text{d.p.}$$

- how to **estimate** the answers to calculations, for example

$$64 \times 18 \text{ is approximately } 60 \times 20 = 1200$$
$$38 \div 0.23 \text{ is approximately } 40 \div 0.2 = 200$$

Exercise 1A

1 Explain the difference between a natural number and an integer.
2 Calculate the answer to each of the following.
   a $(+8) + (-3)$ **b** $(-5) + (+2)$ **c** $(-4) + (-5)$
   d $(+3) - (+2)$ **e** $(+4) - (+8)$ **f** $(-8) - (-6)$
3 Calculate the answer to each of the following.
   a $(+8) \times (-4)$ **b** $(-7) \times (+9)$ **c** $(-5) \times (-9)$
   d $(-36) \div (+6)$ **e** $(+63) \div (-9)$ **f** $(-49) \div (-7)$
4 If $xy = 24$, copy and complete the table below.

x	+8	+6	+4	+2	−2	−4	−6	−8
y								

5 Round each of the following numbers to the degree of accuracy shown in brackets.
   a 2841   (nearest 100) **b** 7286   (nearest 10)   **c** 48 756   (nearest 1000)   **d** 951   (nearest 100)
6 Round the following numbers to one decimal place.
   a 3.84   **b** 6.792   **c** 0.8526   **d** 1.5849
7 Round the following numbers to one significant figure.
   a 3.84   **b** 6.792   **c** 0.7765   **d** 9.624
8 1 mile is 1760 yards. Estimate the number of yards in 11.5 miles.
9 Estimate the area of the figure below.

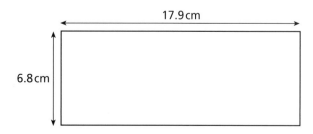

10 Estimate the area of the figure below.

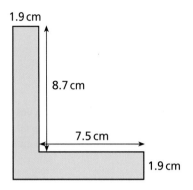

1.9 cm

8.7 cm

7.5 cm

1.9 cm

Exercise 1B

1 Write each of the following as a product of prime factors.

    a 60   **b** 63   **c** 64

2 Find the highest common factor of each of the following pairs of numbers.

    a 16, 36   **b** 42, 68   **c** 24, 48

3 Find the lowest common multiple of each of the following pairs of numbers.

    a 4, 7   **b** 6, 10   **c** 12, 18

4 Estimate the area of the figure below.

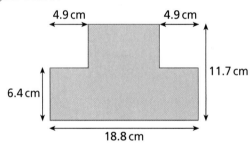

4.9 cm   4.9 cm

11.7 cm

6.4 cm

18.8 cm

5 A cuboid's dimensions are given as 3.9 m by 2.5 m by 3.2 m. Estimate its volume.

6 Estimate the volume of the figure below.

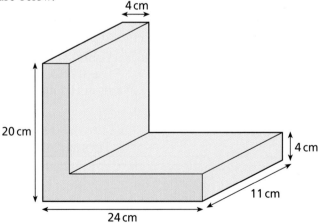

4 cm

20 cm

4 cm

11 cm

24 cm

7 1 foot is approximately 30 cm, 1 yard is 3 feet. Estimate the number of metres in 1000 yards.

8 Estimate the volume of the room you are in. If possible, measure the dimensions of the room to check your answer.

Exercise 1C

The grid below shows a 100 square. A 2×2 square with the numbers 16, 17, 26 and 27 has been highlighted on it.

1	2	3	4	5	6	7	8	9	10
11	12	13	14	15	16	17	18	19	20
21	22	23	24	25	26	27	28	29	30
31	32	33	34	35	36	37	38	39	40
41	42	43	44	45	46	47	48	49	50
51	52	53	54	55	56	57	58	59	60
61	62	63	64	65	66	67	68	69	70
71	72	73	74	75	76	77	78	79	80
81	82	83	84	85	86	87	88	89	90
91	92	93	94	95	96	97	98	99	100

- Using the 2×2 square, add together diagonally opposite corners, i.e.

$$16 + 27 =$$
$$17 + 26 =$$

What do you notice?
- Try other 2×2 squares on the grid, what do you notice? Can you explain why this happens?
- Try adding together opposite corners of 3×3 or 4×4 squares. What happens?
- Investigate patterns involving opposite corners of squares on the grid.
- What happens if you look at opposite corners of rectangles on the grid?

Exercise 1D

You will need:
• computer with a spreadsheet package installed

It is possible to check whether numbers are prime by using a spreadsheet. Below is an example from a spreadsheet.

	A	B	C	D	E
1		**Prime Numbers Check**			
2					
3		Divided by 2	**19.5**		
4		Divided by 3	**13**		
5	**39**	Divided by 5	**7.8**		
6		Divided by 7	**5.571429**		
7		Divided by 11	**3.545455**		
8		Divided by 13	**3**		
9					

Create your own spreadsheet as explained below.

• In cell A5 enter the number to be checked.
• In cell C3 enter a formula to divide the number in A5 by 2.
• In cell C4 enter a formula to divide the number in A5 by 3.
• In cell C5 enter a formula to divide the number in A5 by 5.
• In cell C6 enter a formula to divide the number in A5 by 7.
• Continue for cells C7 and C8.

1 How can you tell from the example above that 39 is not prime?
2 Enter the number 91 in cell A5. How can you decide whether 91 is prime or not?
3 Use your spreadsheet to check the answers you got in exercise 1.6, question 1.

Exercise 1E

You will need:
• encyclopaedia (book or CD-ROM)

or
• computer with internet access

Use an internet browser or encyclopaedia to find:
a what is meant by a Mersenne prime number,
b the biggest prime number known to date.

2 Powers and roots

The Chinese have studied mathematics for many hundreds of years. Some of the world's oldest surviving books on mathematics originate from China. A Chinese mathematician called Wang Xiaotong wrote a book called *The Continuation of Ancient Mathematics* in about 600 AD. In one of its chapters he wrote about calculations involving squares, square roots, cubes and cube roots.

Note that this book was a continuation of an earlier work. It is therefore possible that the work you will be doing in this chapter has echoes of work done in China nearly two thousand years ago!

Squares

The pattern sequence below is made up of 1 cm squares.

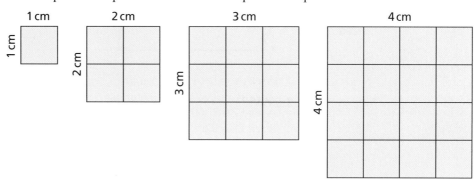

The 1 cm × 1 cm square contains *one* 1 cm square.
The 2 cm × 2 cm square contains *four* 1 cm squares.
The 3 cm × 3 cm square contains *nine* 1 cm squares.
The 4 cm × 4 cm square contains *sixteen* 1 cm squares.

The numbers 1, 4, 9, 16 are known as **square numbers**, and are made by multiplying an integer (whole number) by itself. For example

$8 \times 8 = 64$, therefore 64 is a square number
$2.3 \times 2.3 = 5.29$ 5.29 is *not* a square number as 2.3 is not an integer

Multiplying a number by itself is known as **squaring** the number. For example

8 squared is 8×8
2.3 squared is 2.3×2.3

There is a shorthand notation for writing a number squared. It involves using **indices**. For example

8 squared is written as 8^2
2.3 squared is written as 2.3^2

Using a calculator

The squared button on a calculator usually looks like $\boxed{x^2}$.

Example　Use your calculator to evaluate 17^2.

$\boxed{1}$ $\boxed{7}$ $\boxed{x^2}$ giving an answer of 289

Calculations without a calculator

Calculating the square of a number can be carried out using long multiplication or by visualising the problem diagrammatically.

Example   Using an appropriate diagram, evaluate 1.6^2.

Draw a square of side length 1.6 units.

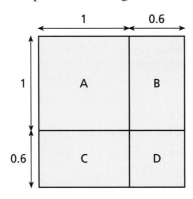

Area A $= 1 \times 1$ $= 1$ units^2
Area B $= 1 \times 0.6$ $= 0.6$ units^2
Area C $= 1 \times 0.6$ $= 0.6$ units^2
Area D $= 0.6 \times 0.6 = 0.36$ units^2
Total $= 2.56$ units^2

Exercise 2.1

1 Calculate how many squares with side of length 1 cm there would be in squares of side:
  a 3 cm   **b** 5 cm   **c** 8 cm   **d** 10 cm
  e 11 cm   **f** 12 cm   **g** 15 cm   **h** 20 cm

2 By drawing appropriate diagrams, calculate the area of squares with sides of the following lengths:
  a 2.1 cm   **b** 3.1 cm   **c** 1.2 cm
  d 5.2 cm   **e** 6.3 cm   **f** 0.7 cm

3 Using long multiplication, work out the area of squares with sides of the following lengths:
  a 2.4 cm   **b** 6.2 cm   **c** 4.6 cm
  d 7.5 cm   **e** 13.2 cm   **f** 23.8 cm

4 Check your answers to questions 2 and 3 by using the $\boxed{x^2}$ key on your calculator.

5 **a** Copy and complete the table below for the
    equation $y = x^2$.

You will need:
• squared/graph paper

x	1	2	3	4	5	6	7	8
y			9					64

  b Plot the graph of $y = x^2$.
  c Use your graph to *estimate* the value of each of the following.
    i) 2.5^2 ii) 3.5^2 iii) 7.2^2 iv) 5.3^2

6 Check your answers to question 5 by using the $\boxed{x^2}$ key on your calculator.

Square roots

Remember:
To find the square root of a number, find the number which, when squared, gives the starting number. For example, the square root of 81 is 9, because 9 × 9 = 81.

The **inverse** (opposite) operation to addition is subtraction; the inverse operation to multiplication is division. Squaring also has an inverse operation: taking the **square root**.

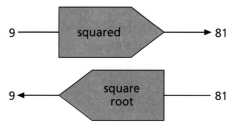

All scientific calculators are able to work out the square root of a number by using the $\boxed{\sqrt{}}$ key.

Examples

Using your calculator, work out $\sqrt{729}$.

$\boxed{\sqrt{}}\ \boxed{7}\ \boxed{2}\ \boxed{9}\ \boxed{=}\ 27$

With some scientific calculators (particularly older ones), the calculation needs to be entered differently. Check how yours works.

Without using a calculator, evaluate $\sqrt{0.36}$.

0.36 can be written as a fraction $\frac{36}{100}$.

$$\sqrt{0.36} = \sqrt{\frac{36}{100}} = \frac{\sqrt{36}}{\sqrt{100}} = \frac{6}{10}$$

$\frac{6}{10} = 0.6$

Therefore $\sqrt{0.36} = 0.6$.

Exercise 2.2

1 *Without using a calculator*, evaluate the following.

  a $\sqrt{25}$ **b** $\sqrt{9}$ **c** $\sqrt{121}$

  d $\sqrt{169}$ **e** $\sqrt{0.01}$ **f** $\sqrt{0.09}$

2 Use the $\boxed{\sqrt{}}$ key on your calculator to check your answers to question 1 above.

3 Evaluate the square root of the following fractions *without the use of a calculator*.

  a $\sqrt{\frac{1}{9}}$ **b** $\sqrt{\frac{1}{49}}$ **c** $\sqrt{\frac{4}{9}}$

  d $\sqrt{\frac{9}{100}}$ **e** $\sqrt{\frac{25}{36}}$ **f** $\sqrt{\frac{49}{81}}$

4 a Copy and complete the table below for the equation $y = \sqrt{x}$.

x	0	1	4	9	16	25	36	49	64	81	100
y											

  b Plot a graph from your table of results above.

  c Use your graph to *estimate* the values of the following square roots:

    i) $\sqrt{70}$ ii) $\sqrt{40}$ iii) $\sqrt{35}$ iv) $\sqrt{55}$ v) $\sqrt{12}$ vi) $\sqrt{95}$

You will need:
• squared/graph paper

5 Check your answers to question 4 using a calculator.

Negative square roots

So far we have only looked at positive square roots. For example, as $5 \times 5 = 25$ we can say that $\sqrt{25} = 5$. However $(-5) \times (-5) = 25$ also, therefore $\sqrt{25}$ can also be -5.

Therefore it is important to realise that taking the square root of a positive number will give two possible values, one positive and one negative, each with the same **magnitude** (same size).

Examples   What are the square roots of 49?

$\sqrt{49} = +7$ and -7.
This is more commonly written as $\sqrt{49} = \pm 7$

....

Evaluate $\sqrt{\frac{4}{25}}$.

$$\sqrt{\frac{4}{25}} = \frac{\sqrt{4}}{\sqrt{25}} = \pm \frac{2}{5}$$

Estimating square roots

We have looked at ways of calculating square roots exactly (using a calculator) and approximately (using a graph). It is also possible to estimate square roots mentally.

Examples   Estimate the value of $\sqrt{18}$.

Consider the two number lines below. The top one is an integer line, the bottom one is a corresponding square root line.

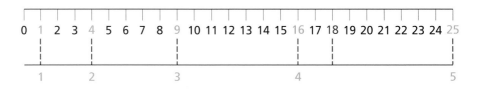

As 18 lies between the square numbers 16 and 25, its square root must lie between 4 and 5. Therefore an estimate for $\sqrt{18}$ is ± 4.2.

....

Estimate the value of $\sqrt{92}$.

As 92 lies between the square numbers 81 and 100, its square root must lie between 9 and 10.
Therefore $\sqrt{92}$ is approximately ± 9.6.

Exercise 2.3

1 Evaluate each of the following, giving all possible solutions.

a $\sqrt{81}$ **b** $\sqrt{169}$ **c** $\sqrt{\frac{1}{36}}$ **d** 11^2 **e** $(-15)^2$ **f** $(-\frac{3}{5})^2$

2 a Copy and complete the following table of values for $y = x^2$.

x	−5	−4	−3	−2	−1	0	1	2	3	4	5
y							1			16	

b Use your table to plot a graph of $y = x^2$ for the range shown.

c Show on your graph how to estimate the following:

i) $\sqrt{19}$ ii) $(−3.7)^2$

> **You will need:**
> • squared/graph paper

3 Estimate the answer to each of the following. Show your working clearly.

a $\sqrt{40}$ **b** $\sqrt{70}$ **c** $\sqrt{150}$ **d** 5.8^2 **e** 3.3^2

4 A carpet layer is laying a carpet in a square room. The area of the room is $46\,\text{m}^2$.

a Estimate the length of each side of the room.

b Explain why there is not a $\pm$ solution to this problem.

5 We have seen that $\sqrt{49} = \pm 7$. Is there a solution to $\sqrt{-49}$? Explain your answer clearly.

Cubes

The pattern sequence below is made up of 1 cm cubes.

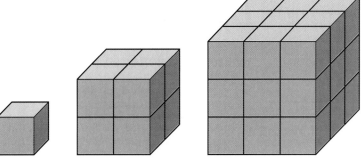

The $1\,\text{cm} \times 1\,\text{cm} \times 1\,\text{cm}$ cube contains *one* 1 cm cube.
The $2\,\text{cm} \times 2\,\text{cm} \times 2\,\text{cm}$ cube contains *eight* 1 cm cubes.
The $3\,\text{cm} \times 3\,\text{cm} \times 3\,\text{cm}$ cube contains *twenty-seven* 1 cm cubes.

The numbers 1, 8 and 27 are known as **cube numbers**, and are made by multiplying an integer by itself three times. For example

$$5 \times 5 \times 5 = 125, \text{ therefore } 125 \text{ is a cube number}$$

Multiplying a number by itself three times is known as **cubing** a number. This, as with squaring, can also be written using indices. For example

$$5 \times 5 \times 5 = 5^3$$

Using a calculator

> Your calculator may have an x^y key. It does the same job.

Few calculators have a specific $\boxed{x^3}$ key. However, all scientific calculators have an indices key. The indices key is as follows: $\boxed{y^x}$. It allows a number to be raised to any power, not just cubed.

Example Using a calculator, evaluate 8^3.

$\boxed{8}$ $\boxed{y^x}$ $\boxed{3}$ $\boxed{=}$ 512

Exercise 2.4

1 How many $1\,cm \times 1\,cm \times 1\,cm$ cubes would make up cubes with sides of the following lengths?

    a $4\,cm$ **b** $6\,cm$ **c** $10\,cm$ **d** $9\,cm$

2 Using the $\boxed{y^x}$ button on your calculator, evaluate the following.

    a 11^3 **b** 20^3 **c** 2.5^3 **d** 6.2^3

3 a Copy and complete the table below for the equation $y = x^3$.

x	0	1	2	3	4	5
y				27		

You will need:
- squared/graph paper

    b Use your table to plot a graph of $y = x^3$.
    c Use your graph to estimate the value of each of the following.
      i) 3.5^3 ii) 4.2^3 iii) 1.8^3
    d Use your calculator to check your answers to part **c** above.

Remember:
To find the cube root of a number, find the number which, when cubed, gives the starting number. For example, the cube root of 27 is 3, because $3 \times 3 \times 3 = 27$.

Cube roots

In the same way that taking the square root is the inverse of squaring, taking the **cube root** is the inverse of cubing.

    Calculators enable this to be done in two main ways. The cube root key is $\boxed{\sqrt[3]{}}$. If a calculator does not have this key, then the general root key is $\boxed{\sqrt[x]{}}$.

Example   Evaluate $\sqrt[3]{64}$ (i.e. find the number which when multiplied by itself three times gives a result of 64).

$\boxed{\sqrt[3]{}}\;\boxed{6}\;\boxed{4}\;\boxed{=}\;4$ (i.e. $4 \times 4 \times 4 = 64$)

Alternatively:

$\boxed{3}\;\boxed{\sqrt[x]{}}\;\boxed{6}\;\boxed{4}\;\boxed{=}\;4$

> Your calculator may have a $y^{1/x}$ or $x^{1/y}$ key. These do the same job, but you will need to use a sequence such as
>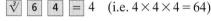
> $\boxed{10}\;\boxed{y^{1/x}}\;\boxed{3}\;\boxed{=}$

> Check to see how your calculator works with this type of calculation.

Exercise 2.5

1 Using a calculator, evaluate the following:

   a $\sqrt[3]{27}$ **b** $\sqrt[3]{216}$ **c** $\sqrt[3]{-343}$ **d** $\sqrt[3]{-0.027}$ **e** $\sqrt[3]{0.001}$

2 Estimate the following calculations. Show your working clearly.

   a 2.8^3 **b** 5.2^3 **c** $\sqrt[3]{90}$ **d** $\sqrt[3]{150}$

3 A water tank, when full, can hold $18\,m^3$ of water. Calculate the length of each side of the tank, if it is in the shape of a cube.

4 A model is shaped as shown below. It consists of two cubes placed one on top of the other.

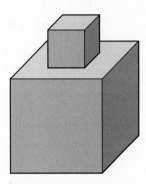

   The length of each side of the smaller cube is $4\,cm$. If the total volume of both cubes is $793\,cm^3$, work out the length of each side of the larger cube.

5 The cuboid shown below has the following characteristics:

- the height is equal to the width
- the length is twice the width.

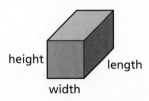

height   length   width

   If the total volume is $1024\,cm^3$, calculate the dimensions of the cuboid.

6 a Copy and complete the table below for the equation $y = x^3$.

x	−5	−4	−3	−2	−1	0	1	2	3	4	5
y								8			

   b Using the table of results above, plot a graph of $y = x^3$ for the range shown.

   c What can you deduce from the shape of the graph about how many solutions there are to the cube root of a number?

   d Use the graph to estimate the value of each of the following:

     i) 4.1^3 ii) $(-2.8)^3$ iii) $\sqrt[3]{80}$ iv) $\sqrt[3]{-50}$

You will need:
- squared/graph paper

SUMMARY

By the time you have completed this chapter you should know:

■ that a whole number multiplied by itself gives a **square number**, for example

$$3 \times 3 = 9$$

■ how to **square** a number

$$1 \times 1 = 1 \qquad 2 \times 2 = 4 \qquad 3 \times 3 = 9$$

■ how to estimate squares from the graph of $y = x^2$
■ that the **inverse** (opposite) of squaring is taking the **square root**
■ how to use the square key $\boxed{x^2}$ and square root key $\boxed{\sqrt{}}$ on your calculator
■ that the square root of a positive number will give two solutions, one positive and one negative; this can be written using the $\pm$ notation, for example

$$\sqrt{64} = \pm 8$$

■ that a **cube number** is the result of multiplying a number by itself three times, for example

$$3 \times 3 \times 3 = 27$$

■ how to **cube** and take the **cube root** of a number, using a calculator if necessary
■ that the cube root of a number will only give one solution, for example

$$\sqrt[3]{27} = 3$$

Exercise 2A

1 Explain in your own words the difference between the **square** and the **square root** of a number.
2 Use your calculator to evaluate the following.
   a $\sqrt{400}$ **b** 7.7^2
3 a Copy and complete the table below for $y = x^3$.

x	−4	−3	−2	−1	0	1	2	3	4
y									

b Use your table to estimate the answer to each of the following.
   i) 2.5^3 ii) $(-3.3)^3$ iii) $\sqrt[3]{52}$ iv) $\sqrt[3]{-30}$

4 The square pattern shown has a total area of $185\,\text{cm}^2$. If the side length of each of the smaller squares is $4\,\text{cm}$, calculate the side length of the larger square.

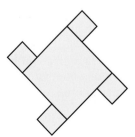

5 The diagram shows a cubical tank filled with water. It has a small cube suspended within it. The side length of the smaller cube is 4 cm. The volume of the water remaining when the smaller cube is removed is 448 cm^3.

Calculate the side length of the larger cube.

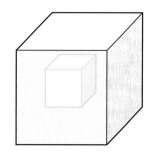

Exercise 2B

1 Explain in your own words, the difference between the **cube** and the **cube root** of a number.

2 Explain, using an appropriate graph of $y = x^2$, why the square root of a positive number will have two solutions.

3 Estimate the following calculations. Show your working clearly.

   a 5.5^2 **b** $\sqrt[3]{200}$

4 Use your calculator to evaluate the following:

   a $(-4.4)^3$ **b** $\sqrt[3]{-1331}$ **c** $5^3 + 6^2$

5 Four right-angled isosceles triangles are joined to form a square as shown. If the area of each triangle is 36 cm^2, calculate the length of each side of the square.

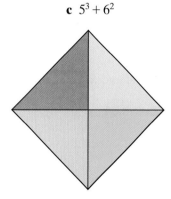

6 A cube has a volume of 3375 cm^3.

   a Calculate the length of each of its edges.

   b Calculate its total surface area.

Exercise 2C

You will need:
• squared paper

An ordinary chessboard is an 8×8 board consisting of 64 small squares, each with a dimension 1×1 unit. There are, however, many other squares of different dimensions within it.

• Start with a square 2×2 board. What is the total number of squares?

- What is the total number of squares in a 3×3 'chessboard'?

- By looking at 4×4 and 5×5 'chessboards' find a number pattern to help you predict the total number of squares in an ordinary chessboard.
- Extend your rule to find the total number of squares in an $n \times n$ board.
- Investigate the total number of squares to be found in an $m \times n$ rectangular 'chessboard'.

Exercise 2D

You will need:
- computer with spreadsheet package installed

The table below lists values of x from 1 to 6 and the corresponding values of x^3.

x	1	2	3	4	5	6
x^3	1	8	27	64	125	216

It can be deduced that the cube root of 100 must be between 4 and 5.

Use a spreadsheet and trial and improvement to find the cube root of 100 correct to 2 d.p.

Exercise 2E

What is a Rubik's cube? See if you can find out more about the man who gave the cube its name.

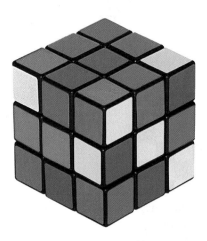
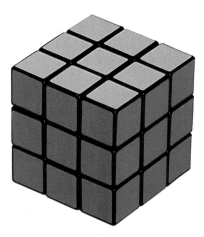

The Rubik's cube is a puzzle which was very popular a number of years ago. See if you can borrow one from your family or friends, and try to solve the puzzle.

3 : Fractions, decimals and percentages

A BBC television series, *The Road to Riches*, dealt with the origin and development of the use of money. Its first programme examined the earliest written records found in the city of Eridu in Mesopotamia (modern Iraq). The records were on tablets like the one shown below.

The Sumerians, as the people of this region were known, used a system of recording value, known as 'cuneiform', five thousand years ago. This writing is now believed to be simply accounts of grain surpluses. This may sound insignificant now, but the change from a hunter–gatherer society to a farming-based society led directly to the kind of sophisticated way of life we have today.

The cuneiform tablets do not record a mathematical system based upon fractions, decimals or percentages, but almost certainly led to the system we use today.

Fractions

Fractions deal with a part of a whole. In its basic form, a fraction has two parts: the **numerator** and the **denominator**.

The numerator is the number above the line; the denominator is the number below the line, i.e.

$$\frac{\text{numerator}}{\text{denominator}}$$

Both the numerator and denominator must be **integers** (whole numbers).

Fractions of a quantity

To work out a fraction of an amount, look at the fraction itself for help. The denominator tells us how many equal parts the amount is split in to; the numerator indicates how many of the parts are being used.

Examples   A teenager spends on average $\frac{2}{5}$ of his pocket money on clothes. If his pocket money is £20 a week, calculate how much he spends on clothes per week.

$\frac{1}{5}$ of £20 is worked out as follows: £20 ÷ 5 = £4.
$\frac{2}{5}$ of £20 is twice as much as $\frac{1}{5}$.
Therefore $\frac{2}{5}$ of £20 is 2 × £4 = £8.
So the teenager spends on average £8 on clothes per week.

A magazine has 56 pages. Of those, $\frac{7}{8}$ contain pictures. Calculate the number of pages with pictures.

$\frac{1}{8}$ of 56 is 56 ÷ 8 = 7.
$\frac{7}{8}$ of 56 is seven times as much as $\frac{1}{8}$.
Therefore $\frac{7}{8}$ of 56 is 7 × 7 = 49.
The magazine has 49 pages with pictures.

Exercise 3.1

1 Evaluate the following amounts.
  a $\frac{1}{4}$ of £6.20   **b** $\frac{1}{8}$ of £25.68   **c** $\frac{4}{9}$ of £127.89   **d** $\frac{3}{7}$ of £2884.35
2 **a** $\frac{7}{12}$ of the pupils in a school are girls. If the school has 1044 pupils, calculate the number of girls.
  b What fraction of the number of pupils in the school are boys?
3 A paint mixture is made up of $\frac{3}{20}$ of red paint, $\frac{9}{20}$ of blue paint and the remainder of white paint.
  a What fraction of the mixture is white paint?
  b If the paint is sold in 5-litre containers, how many litres of red paint are needed for each container?
  c How many litres of blue paint are needed for each 5-litre container?
4 A boy earns £240 a week as a trainee manager. $\frac{1}{5}$ of his earnings are taken in tax. He spends $\frac{1}{4}$ on clothes, $\frac{3}{8}$ on going out with friends and pays his parents £35 a week in rent. The rest he saves.
  a How much tax does he pay?
  b What fraction of his earnings does he pay in rent?
  c How much does he save per week?

Equivalent fractions

Although fractions may look different from each other, sometimes they are worth the same. If fractions are worth the same, they are known as **equivalent fractions**.

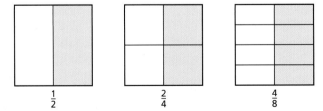

$\frac{1}{2}$ $\frac{2}{4}$ $\frac{4}{8}$

From the diagrams above, the same proportion of each square has been shaded. Therefore we can conclude that the three fractions $\frac{1}{2}$, $\frac{2}{4}$ and $\frac{4}{8}$ are equivalent.

   To determine whether fractions are equivalent, we need to **simplify** them. To simplify a fraction, divide both the numerator and the denominator by their **highest common factor** (the biggest number that goes into both of them exactly).

Examples

Write $\frac{12}{42}$ as a fraction in its simplest form.

> **Remember:**
> *When simplifying, you are making the numerator and denominator as small as possible.*

The highest common factor of 12 and 42 is 6, i.e. 6 is the highest number which goes into both 12 and 42 exactly.

$$12 \div 6 = 2$$
$$42 \div 6 = 7$$

Therefore $\frac{12}{42}$ is simplified to $\frac{2}{7}$.

Which of the following three fractions are equivalent?

$$\frac{4}{18} \qquad \frac{5}{20} \qquad \frac{6}{27}$$

To find which of the three fractions are equivalent to each other, we need to simplify them.

$$\frac{4}{18} = \frac{2}{9} \quad \text{(2 is the highest common factor)}$$
$$\frac{5}{20} = \frac{1}{4} \quad \text{(5 is the highest common factor)}$$
$$\frac{6}{27} = \frac{2}{9} \quad \text{(3 is the highest common factor)}$$

From the above simplifications we can see that $\frac{4}{18}$ and $\frac{6}{27}$ are equivalent.

Exercise 3.2

1 Write each of the following fractions in its simplest form.

a $\frac{2}{10}$ **b** $\frac{3}{27}$ **c** $\frac{9}{48}$ **d** $\frac{48}{56}$ **e** $\frac{34}{85}$ **f** $\frac{65}{104}$

2 In each of the following, determine which of the fractions given are equivalent. Show your working clearly.

a $\frac{3}{4}$ $\frac{5}{7}$ $\frac{15}{21}$ **b** $\frac{5}{8}$ $\frac{20}{32}$ $\frac{18}{30}$ **c** $\frac{4}{9}$ $\frac{16}{36}$ $\frac{28}{63}$ **d** $\frac{3}{10}$ $\frac{9}{30}$ $\frac{6}{24}$ $\frac{15}{50}$ **e** $\frac{6}{32}$ $\frac{4}{16}$ $\frac{3}{16}$ $\frac{15}{80}$ **f** $\frac{3}{8}$ $\frac{2}{10}$ $\frac{6}{15}$ $\frac{4}{22}$

3 Using diagrams, show that each of the following pairs of fractions is equivalent.

You will need:
• squared paper

a $\frac{1}{3} = \frac{3}{9}$ **b** $\frac{2}{5} = \frac{8}{20}$ **c** $\frac{3}{4} = \frac{9}{12}$ **d** $\frac{5}{8} = \frac{15}{24}$

Ordering fractions

Unless fractions are written with the same denominator, it can be difficult, just by looking at them, to see which fraction is the bigger and which is the smaller. For example, with the fractions $\frac{5}{12}$ and $\frac{7}{12}$, it is easy to see that $\frac{7}{12}$ is the larger of the two fractions. However, with $\frac{4}{9}$ and $\frac{5}{12}$, it is not quite so straightforward. To compare and order fractions, it is usual to write them as equivalent fractions with the same denominator.

Example

Write the following fractions in ascending **order of magnitude**.

> **Remember:**
> *magnitude = size*

$$\frac{3}{7} \qquad \frac{8}{21} \qquad \frac{2}{6}$$

Looking at the denominators we can identify 42 as being the **lowest common multiple** of 7, 21 and 6. Therefore writing each of the fractions as equivalent fractions, with a denominator of 42, gives:

> **Remember:**
> *The **lowest common multiple** is the smallest number into which all the denominators will go.*

$$\frac{3}{7} = \frac{18}{42}$$
$$\frac{8}{21} = \frac{16}{42}$$
$$\frac{2}{6} = \frac{14}{42}$$

Therefore the fractions in ascending order are: $\frac{2}{6}, \frac{8}{21}, \frac{3}{7}$.

Exercise 3.3

1 Which of the following statements are true and which are false?

a $\frac{2}{4} > \frac{1}{3}$ **b** $\frac{5}{9} < \frac{6}{10}$ **c** $\frac{4}{12} = \frac{1}{3}$

d $\frac{7}{8} > \frac{8}{10}$ **e** $\frac{8}{15} < \frac{3}{5}$ **f** $\frac{5}{8} = \frac{54}{88}$

2 Write each of the following sets of fractions in ascending order of magnitude.

a $\frac{5}{9}$ $\frac{2}{3}$ $\frac{3}{6}$ **b** $\frac{3}{7}$ $\frac{2}{6}$ $\frac{10}{84}$ **c** $\frac{6}{13}$ $\frac{1}{3}$ $\frac{10}{26}$ **d** $\frac{14}{15}$ $\frac{27}{30}$ $\frac{37}{40}$

3 In a tennis match, player A managed to get $\frac{26}{50}$ of her first serves in. Player B managed to get $\frac{32}{60}$ of her first serves in. Which player got a higher proportion of her first serves in?

4 Two neighbouring schools hold their school play on the same evening. At one school, $\frac{7}{20}$ of the audience are children. At the other, $\frac{9}{25}$ of the audience are children. Which school has the higher proportion of children in the audience?

5 A tropical fruit juice states on the label that it is made up of the following fractions of fruit:

     $\frac{1}{4}$ orange $\frac{2}{7}$ mango $\frac{3}{10}$ passion fruit $\frac{1}{8}$ pineapple $\frac{11}{280}$ grape

Write down the fractions in ascending order of magnitude.

Calculations with fractions

To carry out calculations with fractions, there are certain rules of which you need to be aware.

Addition and subtraction

To add or subtract fractions with the same denominator is relatively straightforward, for example:

$$\frac{1}{8} + \frac{3}{8} = \frac{4}{8} = \frac{1}{2}$$

Visually this can be shown as follows:

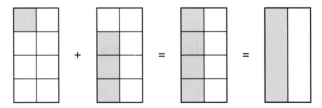

Simply add the numerators together and keep the denominator as it is.

However, adding or subtracting fractions with different denominators is a little less straightforward. For example

$$\frac{1}{4} + \frac{2}{5}$$

In order to do this, both fractions need to be converted into equivalent fractions with a common denominator. The lowest common multiple of both denominators is 20 (i.e. 20 is the smallest number that both 4 and 5 go in to). Therefore we should aim to find equivalent fractions to those given, with 20 as a denominator.

$$\frac{1}{4} = \frac{5}{20} \quad \text{and} \quad \frac{2}{5} = \frac{8}{20}$$

Therefore $\frac{1}{4}+\frac{2}{5}$ is the same as $\frac{5}{20}+\frac{8}{20}$. Visually this can be shown as follows:

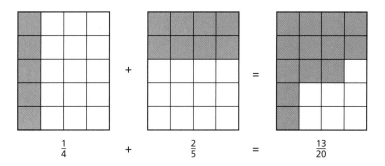

$$\frac{1}{4} \qquad + \qquad \frac{2}{5} \qquad = \qquad \frac{13}{20}$$

Note that, with subtraction, it is also necessary to work with fractions with a common denominator.

Multiplication

Multiplying fractions is relatively straightforward. For example

$$\tfrac{1}{3}\times\tfrac{3}{5}$$

This can be read as $\frac{1}{3}$ of $\frac{3}{5}$. To represent this visually, once again find the lowest common multiple of the denominators (i.e. 15). Shade $\frac{3}{5}$ of the pattern, which is equivalent to $\frac{9}{15}$.

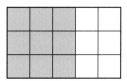

Shade $\frac{1}{3}$ of the $\frac{3}{5}$, giving a fraction of $\frac{3}{15}$ of the original pattern.

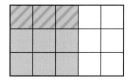

Therefore $\frac{1}{3}$ of $\frac{3}{5}$ is equal to $\frac{3}{15}$.

Doing this is equivalent to simply multiplying the numerators together and multiplying the denominators together to give the answer.

$$\tfrac{1}{3}\times\tfrac{3}{5}=\tfrac{3}{15}=\tfrac{1}{5}$$

Division

To understand division involving fractions, it is easiest to look at the problem visually. For example

$\frac{2}{3} \div \frac{1}{4}$ can be read as 'how many quarters go in to $\frac{2}{3}$?'

$\frac{1}{4}$ is equivalent to $\frac{3}{12}$

$\frac{2}{3}$ is equivalent to $\frac{8}{12}$

Therefore we are trying to see how many $\frac{3}{12}$ go into $\frac{8}{12}$:

This can be seen to give $2\frac{2}{3}$.

Another way of carrying out a division involving fractions is to transform it into a multiplication. For example

$\frac{2}{3} \div \frac{1}{4}$ is the same as $\frac{2}{3} \times \frac{4}{1} = \frac{8}{3} = 2\frac{2}{3}$

> **Remember:**
> *The reciprocal of 7 is $\frac{1}{7}$.*
> *The reciprocal of $\frac{3}{2} = \frac{2}{3}$.*

A division can therefore be changed to a multiplication by the **reciprocal**.

Exercise 3.4

1 Add the following fractions. Simplify your answers where possible.
   a $\frac{1}{7} + \frac{4}{7}$ **b** $\frac{1}{9} + \frac{5}{9}$ **c** $\frac{4}{15} + \frac{6}{15}$ **d** $\frac{5}{24} + \frac{7}{24}$

2 Work out the answer to each of the following calculations. Show your working clearly and simplify your answers where possible.
   a $\frac{2}{5} + \frac{1}{6}$ **b** $\frac{7}{12} + \frac{1}{5}$ **c** $\frac{9}{14} - \frac{2}{7}$ **d** $\frac{3}{13} - \frac{3}{26}$
   e $\frac{1}{8} + \frac{5}{16} - \frac{5}{24}$ **f** $\frac{13}{18} - \frac{8}{9} + \frac{1}{6}$

3 Sadiq spends $\frac{1}{5}$ of his earnings on his mortgage. He saves $\frac{2}{7}$ of his earnings. What fraction of his earnings is left?

4 A farmer uses five out of seven equal strips of his land for cereal crops and $\frac{1}{8}$ of his land for root vegetables. What fraction of his land is available for other uses?

5 Work out the following calculations, simplifying your answers where possible.
   a $\frac{1}{6} \times \frac{3}{8}$ **b** $\frac{2}{3} \times \frac{5}{12}$ **c** $\frac{3}{10} \times \frac{5}{7}$ **d** $\frac{1}{3} \times \frac{1}{4} \times \frac{1}{5}$

6 Work out the following, showing your working clearly and simplifying your answers where possible.
   a $\frac{5}{9} \div \frac{1}{3}$ **b** $\frac{11}{14} \div \frac{3}{21}$ **c** $\frac{9}{16} \div \frac{2}{5} \times \frac{3}{4}$ **d** $\frac{1}{2} \div \frac{1}{3} \times \frac{1}{4}$

Use of a calculator

Most scientific calculators have a fraction button. In addition to using it to carry out calculations with fractions, it enables you to simplify fractions, compare magnitudes and convert fractions to decimals.

The fraction button usually looks like this $\boxed{a^b/_c}$.

Examples   Use the fraction button on a calculator to simplify $\frac{27}{33}$.

This gives a solution of $\frac{9}{11}$.

A 5 km race takes a runner 25 minutes to complete. For $\frac{3}{5}$ of the time he is in first place and for $\frac{1}{6}$ of the time he is in second place.
a Calculate how long the runner spent in first place.
b How long did the runner spend in second place?

a Use the following key sequence.

$\boxed{3}\ \boxed{a^b/_c}\ \boxed{5}\ \boxed{\times}\ \boxed{2}\ \boxed{5}\ \boxed{=}$

The runner spent 15 minutes in first place.
b Use the following key sequence.

$\boxed{1}\ \boxed{a^b/_c}\ \boxed{6}\ \boxed{\times}\ \boxed{2}\ \boxed{5}\ \boxed{=}$

Here the calculator gives an answer of $4\frac{1}{6}$ minutes.
$\frac{1}{6}$ of a minute is 10 seconds.
Therefore the runner spent 4 minutes and 10 seconds in second place.

Exercise 3.5

In each of the following questions, use the fraction button on your calculator if you need to.

1 On a typical school day, a pupil spends 6 hours in lessons.
   a What fraction of the school day is spent in lessons?
   b Write this fraction in its simplest form.
2 One day Maria spends $\frac{6}{20}$ of the day asleep. Calculate in hours and minutes the amount of time she spends asleep.
3 Paul sat a $2\frac{1}{2}$ hour exam. If he finished $\frac{13}{15}$ of the way through the exam, calculate in minutes the amount of time he had spare at the end.
4 A marathon runner took 2 hours 15 minutes to complete the race. During that time he spent 50 minutes in the lead. Write in its simplest form the fraction of time he spent in the lead.
5 Michael worked out that, for 17 out of every 20 minutes he spends on his mobile phone, he is talking to friends. If he spent 3 hours 45 minutes on the phone, calculate how long he spent talking to friends.
6 How many minutes are represented by each of the following fractions of an hour?
   a $\frac{4}{30}$ **b** $\frac{5}{12}$ **c** $\frac{3}{10}$
   d $\frac{4}{5}$ **e** $\frac{11}{15}$ **f** $\frac{6}{45}$
7 Express each of the following number of minutes as a fraction of an hour in its simplest form.
   a 27 min   **b** 14 min   **c** 48 min
   d 35 min   **e** $52\frac{1}{2}$ min   **f** 57 min

Decimals

The Dutch mathematician Simon Stevin (1584–1620) wrote one of the first books on the theory of decimals. He wrote 2.378 as 2 ⓪ 3 ① 7 ② 8. Other ways of writing this at the time were 2<u>378</u> and 2I<u>378</u>.

Decimals are another way of writing parts of a whole number. For example, the fraction $\frac{1}{2}$ can be written as a decimal as 0.5.

To convert a fraction to a decimal, simply divide the numerator by the denominator.

Example   Write $\frac{2}{5}$ as a decimal.

$2 \div 5 = 0.4$.
Therefore $\frac{2}{5}$ is 0.4 as a decimal.

There are different types of decimal. A **terminating decimal** is one in which the numbers after the decimal point come to an end. A **recurring decimal** is one in which the numbers repeat themselves, and would continue to do so forever.

Examples   Convert $\frac{1}{5}$ to a decimal. Decide whether the decimal is terminating or recurring.

$\frac{1}{5} = 0.2$.
0.2 is a terminating decimal.

Convert $\frac{2}{3}$ to a decimal. Decide whether the decimal is terminating or recurring.

$\frac{2}{3} = 0.66666666666666\ldots$
This is an example of a recurring decimal. It can be written as $0.\dot{6}$. The 'dot' above the 6 implies that it repeats.

Convert $\frac{12}{99}$ to a decimal. Decide whether the decimal is terminating or recurring.

$\frac{12}{99} = 0.1212121212121212\ldots$
This is another example of a recurring decimal. This can be written as $0.\dot{1}\dot{2}$. The 'dots' in this case indicate that both numbers repeat.

Where three numbers repeat, for example 7.128128..., this is written as $7.\dot{1}2\dot{8}$.

Exercise 3.6

1 Using a calculator (or spreadsheet), copy and complete this table. Divide each numerator by each denominator to convert fractions to decimals.

		numerator											
		1	2	3	4	5	6	7	8	9	10	11	12
denominator	1												
	2												
	3												
	4												
	5												
	6												
	7												
	8												
	9												
	10												
	11												
	12												

By looking at your table from question 1, answer the following questions.

2 Describe the type of fraction which gives a decimal answer greater than 1.

3 Describe the type of fraction which gives an answer of 1.

4 Name three fractions that give terminating decimals.

5 Do fractions with a '7' as the denominator give recurring decimals? Explain your answer clearly.

6 Describe the different types of recurring decimal that you see.

Changing a decimal to a fraction

Just as fractions can be written as either terminating or recurring decimals, the reverse is also true. All terminating or recurring decimals can be written as fractions. To do this, an understanding of place value is necessary.

Examples Write the decimal 0.6 as a fraction in its simplest form.

By entering 0.6 into a place value table we get:

units	•	tenths
0	•	6

The '6' is worth six tenths. As a fraction 0.6 can therefore be written as $\frac{6}{10}$. $\frac{6}{10}$ can in turn be simplified to $\frac{3}{5}$.

Write the decimal 0.325 as a fraction in its simplest form.

By entering 0.325 into a place value table we get:

units	•	tenths	hundredths	thousandths
0	•	3	2	5

3 tenths, 2 hundredths and 5 thousandths is equivalent to 325 thousandths. As a fraction this can therefore be written as $\frac{325}{1000}$. $\frac{325}{1000}$ can be simplified to $\frac{13}{40}$.

Write the recurring decimal $0.\dot{1}\dot{3}$ as a fraction in its simplest form.

This poses difficulties because the decimal is infinite, and so cannot be entered into a place value table. We can overcome this by turning to algebra.
    Let

$$x = 0.1313131313\ldots$$
$$100x = 13.131313131313\ldots$$

By subtracting one equation from the other, we get:

$$99x = 13$$

therefore $x = \frac{13}{99}$.

> Note: the trick here is to multiply x by a power of ten such that the numbers after the decimal point match those of the original decimal.

Place value tables can also be used to compare the size of two or more decimals.

Example Which of the following two decimals is the larger?

$$0.2 \quad \text{or} \quad 0.18$$

Entering them both into a place value table gives:

units	• tenths	hundredths
0	• 2	0
0	• 1	8

We can see that 0.2 is equivalent to 2 tenths or 20 hundredths, and 0.18 is equivalent to 18 hundredths. Therefore 0.2 is larger than 0.18.

Exercise 3.7

1 Convert each of the following fractions to a decimal.

   a $\frac{1}{20}$ **b** $\frac{3}{15}$ **c** $\frac{7}{28}$

   d $\frac{3}{14}$ **e** $\frac{1}{24}$ **f** $\frac{17}{99}$

2 Convert each of the following decimals to a fraction in its simplest form. Show your working clearly.

   a 0.3   **b** 0.12   **c** 0.625

   d 0.37   **e** 0.2125

3 Convert each of the following recurring decimals to a fraction. Show your working clearly.

   a $0.\dot{6}$ **b** $0.3\dot{7}$ **c** $0.\dot{7}\dot{5}$

   d $0.0\dot{1}$ **e** $0.3\dot{5}\dot{3}$ **f** $0.\dot{2}8571\dot{4}$

4 Write each of the following sets of decimals in ascending order.

   a 0.7   0.55

   b 0.27   0.100

   c 0.625   0.8   0.73

   d 0.303   0.33   0.3003

   e 0.01   0.10   0.101

   f 0.32   0.43   0.403

Calculations with decimals

We carry out calculations with decimals every day when dealing with money. Although most of those calculations are now carried out using a calculator or computer, it is nevertheless important to have an understanding of how answers are obtained, least of all to check that you are not out of pocket!

Examples Paul went to a café and bought a sandwich costing £1.85 and a drink costing 73p. Calculate the total cost of his bill.

This sum can be set out in the conventional way, ensuring that place value is taken into account.

$$
\begin{array}{r}
1.85 \\
+\ 0.73 \\
\hline
2.58 \\
\hline
{}_{1}
\end{array}
$$

The total cost of Paul's bill is £2.58.

Maggie went shopping and bought four birthday cards, each priced £1.53. Calculate the total cost of the four cards.

$$
\begin{array}{r}
1.53 \\
\times\quad 4 \\
\hline
6.12 \\
\hline
{}_{2}\ \ {}_{1}
\end{array}
$$

Therefore Maggie's bill came to £6.12.

£135.50 was shared equally between five friends. How much did they each receive?

$$
\begin{array}{r}
2\ \ 7.1\ 0 \\
\hline
5\ \lvert\ \ 1\ 3\ {}^{3}5.5\ 0
\end{array}
$$

Each of the five friends received £27.10.

Exercise 3.8

1 Marco goes shopping for food. The list of items and their prices are given below.

bread	63p
soup	£1.25
coffee	£2.18
sugar	87p
yoghurt	42p

a Calculate the total amount Marco had to pay.
b If he paid with a £10 note, calculate the change he received.

2 A family of two adults and two children decided to visit a theme park for the day. An individual adult ticket costs £19.20 and an individual child ticket costs £16.70.
a Calculate the cost of buying the two adult's tickets and two children's tickets.
b Calculate how much the family would save if, instead of buying individual tickets, they bought a family ticket costing £57.

3 A brother and sister are 1.63 m tall and 94 cm tall respectively. What is the difference in their heights:
a in metres   b in centimetres?

4 Isabel buys two pairs of socks costing £3.67 each and three pairs of trousers costing £34.49 each.
a Calculate the total cost of these items.
b If Isabel had only £42 in her bank account, work out by how much she had overdrawn.

5 Nine people share a lottery prize of £7 365 800. Calculate, to the nearest penny, how much they each receive.

6 A table and six chairs are priced at £335 in a garden centre. If the table costs £120.80, calculate the cost of each chair.

Percentages

Percentages, fractions and decimals are all different ways of representing the same values. The unique feature of percentages is that they are written as values out of 100.

You should already be familiar with most of the percentage equivalents of simple fractions and decimals outlined in the table on the right.

fraction	decimal	percentage
$\frac{1}{2}$	0.5	50%
$\frac{1}{4}$	0.25	25%
$\frac{3}{4}$	0.75	75%
$\frac{1}{8}$	0.125	12.5%
$\frac{3}{8}$	0.375	37.5%
$\frac{5}{8}$	0.625	62.5%
$\frac{7}{8}$	0.875	87.5%
$\frac{1}{10}$	0.1	10%
$\frac{2}{10}$ or $\frac{1}{5}$	0.2	20%
$\frac{3}{10}$	0.3	30%
$\frac{4}{10}$ or $\frac{2}{5}$	0.4	40%
$\frac{6}{10}$ or $\frac{3}{5}$	0.6	60%
$\frac{7}{10}$	0.7	70%
$\frac{8}{10}$ or $\frac{4}{5}$	0.8	80%
$\frac{9}{10}$	0.9	90%

Simple percentages

Examples

Of 100 sheep in a field, 88 are ewes.
a What percentage of sheep are ewes?
b What percentage are not ewes?

a 88 out of 100 are ewes, i.e. 88%.
b 12 out of 100 are not ewes, i.e. 12%.

A gymnast scored the following marks out of 10 from five judges:

$$8.0 \quad 8.2 \quad 7.9 \quad 8.3 \quad 7.6$$

Express these marks as percentages.

$$\frac{8.0}{10} = \frac{80}{100} = 80\% \qquad \frac{8.2}{10} = \frac{82}{100} = 82\% \qquad \frac{7.9}{10} = \frac{79}{100} = 79\%$$

$$\frac{8.3}{10} = \frac{83}{100} = 83\% \qquad \frac{7.6}{10} = \frac{76}{100} = 76\%$$

Convert the following percentages into fractions and decimals.
a 27%   **b** 5%

a $\frac{27}{100} = 0.27$
b $\frac{5}{100} = 0.05$

Exercise 3.9

1 In a survey of 100 cars, 47 were white, 23 were blue and 30 were red. Express each of these numbers as a percentage of the total.
2 $\frac{7}{10}$ of the surface of the Earth is water. Express this as a percentage.
3 There are 200 birds in a flock and 120 of them are female. What percentage of the flock is:
  a female   **b** male?
4 Write these percentages as fractions of 100.
  a 73%   **b** 28%   **c** 10%   **d** 25%
5 Write these fractions as percentages.
  a $\frac{27}{100}$ **b** $\frac{3}{10}$ **c** $\frac{7}{50}$ **d** $\frac{1}{4}$
6 Convert the following percentages to decimals.
  a 39%   **b** 47%   **c** 83%
  d 7%   **e** 2%   **f** 20%
7 Convert the following decimals to percentages.
  a 0.31   **b** 0.67   **c** 0.09
  d 0.05   **e** 0.2   **f** 0.75

Calculating a percentage of a quantity

Examples   Find 25% of 300 m.

25% can be written as 0.25.
$0.25 \times 300\,\text{m} = 75\,\text{m}$.

Find 35% of 280 m.

35% can be written as 0.35.
$0.35 \times 280\,\text{m} = 98\,\text{m}$.

Exercise 3.10

1 Write the percentage equivalent of each of the following fractions.

   a $\frac{1}{4}$ **b** $\frac{2}{3}$ **c** $\frac{5}{8}$

   d $1\frac{4}{5}$ **e** $4\frac{9}{10}$ **f** $3\frac{7}{8}$

2 Write the decimal equivalent of each of the following.

   a $\frac{3}{4}$ **b** 80%   **c** $\frac{1}{5}$

   d 7%   **e** $1\frac{7}{8}$ **f** $\frac{1}{6}$

3 Evaluate:

   a 25% of 80   **b** 80% of 125   **c** 62.5% of 80

   d 30% of 120   **e** 90% of 5   **f** 25% of 30

4 Evaluate:

   a 17% of 50   **b** 50% of 17   **c** 65% of 80

   d 80% of 65   **e** 7% of 250   **f** 26% of 7

5 In a class of 30 students, 20% have black hair, 10% have blond hair and 70% have brown hair. Calculate the number of students with:

   a black hair   **b** blond hair   **c** brown hair

6 A survey conducted among 120 schoolchildren looked at which type of meat they preferred. 55% said they preferred chicken, 15% preferred lamb and 10% preferred pork. The rest were vegetarian. Calculate the number of children in each category.

7 A survey was carried out in a school to see from which ethnic background its students came. Of the 220 students in the school, 65% were English, 20% were Pakistani, 5% were Afro-Caribbean and 10% belonged to other backgrounds. Calculate the number of students from each ethnic background.

8 A shopkeeper keeps a record of the number of items he sells in one day. Of the 150 items he sold, 46% were newspapers, 24% were pens, 12% were books and the remaining 18% were other assorted items. Calculate the number of each item sold.

Expressing one quantity as a percentage of another

To write one quantity as a percentage of another, write the first quantity as a fraction of the second and then multiply by 100.

Example   In an examination a girl gets 69 marks out of 75. Express this result as a percentage.

$\frac{69}{75} \times 100 = 92$

She gets 92%.

Exercise 3.11

1 Express the first quantity as a percentage of the second.
  a 24 out of 50   **b** 46 out of 125   **c** 7 out of 20   **d** 45 out of 90
  e 9 out of 20   **f** 16 out of 40   **g** 13 out of 39   **h** 20 out of 35
2 A hockey team plays 42 matches. It wins 21, draws 14 and loses the rest. Express each of these results as a percentage of the total number of games played.
3 Four candidates stood in an election:

    A received 24 500 votes
    B received 18 200 votes
    C received 16 300 votes
    D received 12 000 votes

Express each of these as a percentage of the total votes cast.
4 A car manufacturer produces 155 000 cars a year. The cars are available for sale in six different colours. The numbers sold of each colour were:

red	55 000
blue	48 000
white	27 500
silver	10 200
green	9 300
black	5 000

Express each of these as a percentage of the total number of cars produced. Give your answers to one decimal place.

Percentage increases and decreases

Examples   A garage increases the price of a truck by 12%. If the original price was £14500, calculate its new price.

The original price represents 100%, therefore the increased price can be written as 112%.

New price = 112% of £14500
= 1.12 × £14500
= £16240

> **Remember:**
> *To change a percentage to a decimal, divide by 100.*

A French doctor has a salary of 3000 euro per month. If her salary increases by 8%, calculate:
a the amount extra she receives a month,
b her new monthly salary.

a Increase = 8% of 3000 euro
= 0.08 × 3000 euro
= 240 euro
b New salary = old salary + increase
= 3000 + 240 euro
= 3240 euro per month

A shop is having a sale. It sells a set of tools costing £130 at a 15% discount. Calculate the sale price of the tools.

The old price represents 100%, therefore the new price can be represented as (100 − 15)% = 85%.

85% of £130 = 0.85 × £130
= £110.50

Exercise 3.12

1 Increase the following by the given percentage. Show your working clearly.
  a 150 by 25%   **b** 230 by 40%   **c** 7000 by 2%   **d** 70 by 250%   **e** 80 by 12.5%   **f** 75 by 62%
2 Decrease the following by the given percentage. Show your working clearly.
  a 120 by 25%   **b** 40 by 5%   **c** 90 by 90%   **d** 1000 by 10%   **e** 80 by 37.5%   **f** 75 by 42%
3 A farmer increases the total yield of grain on his farm by 15%. If his previous yield was 6500 tonnes, what is his present yield?
4 The cost of a computer in a German shop is reduced by 12.5% in a sale. If the computer was priced at 7800 euro, what is its price in the sale?
5 A winter coat is priced at £100. In the sale its price is reduced by 25%.
  a Calculate the sale price of the coat.
  b After the sale, the coat's price is increased by 25%. Calculate the coat's price after the sale.
6 A builder charges a price of £850 for a job. If VAT at 17.5% is added on to the price, calculate the total price of the job.
7 A farmer takes 250 chickens to be sold at a market. In the first hour he sells 8% of his chickens. In the second hour he sells 10% of those he had left.
  a How many chickens has he sold in total?
  b What percentage of the original number did he manage to sell in the two hours?

SUMMARY

By the time you have completed this chapter you should know:

■ how to calculate a fraction of a quantity

$$\tfrac{1}{5} \text{ of } 20 = 20 \div 5$$
$$= 4$$

■ how to identify **equivalent fractions**

$$\tfrac{1}{2} = \tfrac{2}{4} = \tfrac{4}{8}$$

■ how to **simplify** fractions

$$\tfrac{9}{30} = \tfrac{3}{10}$$

■ how to arrange fractions in **order of magnitude**

■ how to carry out calculations with fractions involving addition, subtraction, multiplication and division

■ how to use the fraction button on your calculator $\boxed{a^{b}/_{c}}$

■ that decimals are another way of writing fractions

$$0.4 = \tfrac{4}{10}$$

■ how to convert **terminating** and **recurring decimals** to fractions

■ how to arrange decimals in order of magnitude

■ how to carry out simple calculations involving decimals without the use of a calculator

■ how to convert between fractions, decimals and percentages

■ how to work out a percentage of a quantity

$$65\% \text{ of } 150 = 0.65 \times 150$$
$$= 97.5$$

■ how to express one quantity as a percentage of another

$$30 \text{ out of } 40 = \tfrac{30}{40} \times 100$$
$$= 75\%$$

■ how to carry out calculations involving percentage increases and decreases.

Exercise 3A

1 In a street of 120 houses, $\tfrac{1}{5}$ of the houses have only one occupant, $\tfrac{1}{3}$ have two occupants, $\tfrac{5}{12}$ have three occupants, and the remainder have four or more occupants.

  a Calculate the number of houses with one occupant.

  b Calculate the number of houses with four or more occupants.

2 Simplify each of the following fractions.

  a $\tfrac{15}{35}$ **b** $\tfrac{27}{36}$ **c** $\tfrac{48}{144}$

3 A group of three friends, Richard, Jo and Anna, share £60 between them. Richard has $\tfrac{6}{15}$ of the total, Jo has $\tfrac{5}{12}$ and Anna has the rest.

  a What fraction of the total amount does Anna receive?

  b Write the three fractions in descending order of magnitude.

4 Work out, without using a calculator, the answer to each of the following calculations.

  a $\tfrac{1}{8} + \tfrac{5}{6}$ **b** $\tfrac{7}{12} - \tfrac{2}{5}$ **c** $\tfrac{3}{7} \times \tfrac{4}{9}$ **d** $\tfrac{8}{15} \div \tfrac{3}{5}$

5 Convert the following decimals to fractions.

  a 0.4   **b** $0.1\dot{2}\dot{3}$

6 In 1999 the average price of a house rose by 12%. If the average price of a house was £67 000 in 1998, calculate its 1999 value.

7 Unemployment figures at the end of last month were 867 000. If the number of unemployed has dropped by 8%, calculate the number of unemployed this month.

8 Petrol costs 84p per litre and 63p of this is tax. Calculate the percentage that motorists pay in tax.

Exercise 3B

You will need:
• squared paper

1 Demonstrate using appropriate diagrams why $\frac{4}{14}$ is equivalent to $\frac{6}{21}$.

2 In a class of 30 children, 12 have blonde hair, 8 have brown hair, 6 have black hair and 4 have auburn hair. Write each of the hair colours as fractions of the total number of children in the class. Give each fraction in its simplest form.

3 Without using a calculator, work out the answers to the following calculations.

  a $\frac{3}{10} + \frac{4}{15}$ **b** $\frac{9}{39} - \frac{3}{26}$ **c** $\frac{8}{9} \times \frac{3}{4}$ **d** $\frac{7}{16} \div \frac{3}{4}$

4 Louise gets paid £4.32 per hour. If she works for 7 hours one Saturday, calculate, without using a calculator, the total amount she earns.

5 Vladimir buys the following items at a newsagent:

newspaper	45p
pen	£1.09
birthday card	£1.73
sweets	68p
3 stamps	29p each

  If he pays using a £10 note, calculate the amount of change he receives.

6 A school tuck shop records the percentage of each flavour of crisps that it sells. The percentages are as follows:

ready salted	15%
salt and vinegar	26%
chicken	38%
cheese and onion	12%
other	9%

  If the shop sells 120 packets of crisps, calculate, to the nearest whole number, the number of packets of each flavour that is sold.

7 The value of shares in a mobile phone company rises by 17%.

  a If the value of each share was originally 324p, calculate, to the nearest penny, the new value of each share.

  b How many shares can now be bought with £250?

8 A shop cuts the price of all of its items by 35% during its summer sale. The list of items below gives their *pre-sale* price.

trainers	£42.50
CDs	£13.99
jeans	£38

  If Nadine buys a pair of trainers, three CDs and two pairs of jeans in the sale, calculate how much she will have spent altogether.

Exercise 3C

Investigating a fraction triangle

In the triangle below, a, b, c, d, e and f all represent fractions.

Fraction d = a − b, fraction e = b − c and fraction f = d − e.

$$
\begin{array}{cccc}
\text{a} & \text{b} & \text{c} \\
\text{d} & \text{e} \\
\text{f}
\end{array}
$$

Using this method of construction, copy and complete the fraction triangle below:

$$
\begin{array}{ccccccc}
1 & \frac{1}{2} & \frac{1}{3} & \frac{1}{4} & \frac{1}{5} & \frac{1}{6} & \frac{1}{7} \\
\frac{1}{2} & ? & ? & \frac{1}{20} & ? & ? \\
? & \frac{1}{12} & ? & ? & ? \\
? & ? & ? & ? \\
& & \text{etc.}
\end{array}
$$

Describe the patterns you see in your completed fraction triangle.

Exercise 3D

'Coats 'R' Us' is a factory manufacturing coats. The table below identifies the raw materials needed for each coat and also their cost to the factory.

You will need:
• computer with spreadsheet package installed.

material	amount per coat	cost
cotton	3.8 m	£200 per 100 m roll
lining	2.3 m	£120 per 100 m roll
thread	180 m	£7 per 2000 m reel
buttons	8	£6 per 50
label	1	£25 per 1000
labour	20 min	£4.50 per hour

The factory has received an order from a large retailer for 20 000 coats.

• Set up a spreadsheet to work out how much of each raw material the factory will need to order.
• Using the spreadsheet, calculate the cost of manufacturing one coat.
• The factory wants to sell each coat for 30% more than it costs to manufacture. Calculate how much each coat will cost the retailer.
• If the cost of labour increases by 4%, use the spreadsheet to work out how much each coat will now cost the retailer.

Exercise 3E

Where did our present number system originate?

What were the limitations of the Roman system of numerals?

See if you can discover when and how the zero (or nought) and the decimal point began to be used in European trade and mathematics.

4 : Ratio and proportion

For a civilisation to endure and prosper, it must give its citizens order and fairness. The Chinese civilisation prospered for many centuries, and part of its fairness depended upon mathematics. Zhoubi Zuanjing wrote one of the great books of mathematics, entitled *Nine Chapters of the Mathematician's Art*, more than two thousand years ago. Three chapters were concerned with ratio and proportion, so that, 'rice and other cereals could be planted and distributed in the correct proportion for our needs', and so that 'the ratio of taxes could be levied fairly'.

Fractions and ratio

Equivalent fractions

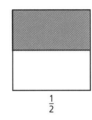 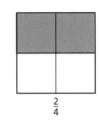 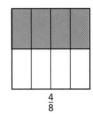

$$\frac{1}{2} \qquad \frac{2}{4} \qquad \frac{4}{8}$$

It should be clear from the diagram above that $\frac{1}{2}$, $\frac{2}{4}$ and $\frac{4}{8}$ are **equivalent fractions**. They are equivalent because each fraction is worth the same amount. Similarly $\frac{1}{3}$, $\frac{2}{6}$ and $\frac{3}{9}$ are equivalent fractions, as are $\frac{1}{5}$, $\frac{10}{50}$ and $\frac{20}{100}$.

Equivalent ratios

Ratios behave in a similar way to fractions. $1:2$ is equivalent to $2:4$ and to $35:70$. In the same way, $15:5$ is equivalent to $3:1$ and $9:3$.

Exercise 4.1

Copy the following sets of equivalent fractions and fill in the blanks.

1 $\dfrac{1}{4} = \dfrac{2}{\rule{1em}{0.4pt}} = \dfrac{\rule{1em}{0.4pt}}{16} = \dfrac{\rule{1em}{0.4pt}}{64} = \dfrac{3}{\rule{1em}{0.4pt}}$

2 $\dfrac{2}{5} = \dfrac{4}{\rule{1em}{0.4pt}} = \dfrac{\rule{1em}{0.4pt}}{20} = \dfrac{\rule{1em}{0.4pt}}{50} = \dfrac{16}{\rule{1em}{0.4pt}}$

3 $\dfrac{3}{8} = \dfrac{6}{\rule{1em}{0.4pt}} = \dfrac{\rule{1em}{0.4pt}}{24} = \dfrac{15}{\rule{1em}{0.4pt}} = \dfrac{\rule{1em}{0.4pt}}{72}$

Copy the following sets of equivalent ratios and fill in the blanks.

4 $4:5 \qquad 8:\rule{1em}{0.4pt} \qquad \rule{1em}{0.4pt}:50 \qquad 12:\rule{1em}{0.4pt}$
5 $7:2 \qquad 14:\rule{1em}{0.4pt} \qquad \rule{1em}{0.4pt}:10 \qquad 49:\rule{1em}{0.4pt}$
6 $8:5 \qquad \rule{1em}{0.4pt}:50 \qquad 32:\rule{1em}{0.4pt} \qquad 4:\rule{1em}{0.4pt}$

Direct proportion

Workers in a pottery factory are paid according to how many plates they produce. The wage paid to them is said to be in **direct proportion** to the number of plates made. As the number of plates made increases so does their wage. Other workers are paid for the number of hours worked. For them the wage paid is in direct proportion to the number of hours worked. There are two main methods for solving problems involving direct proportion: the **ratio method** and the **unitary method**.

Example

A bottling machine fills 500 bottles in 15 minutes. How many bottles will it fill in $1\frac{1}{2}$ hours?

Note. The time units must be the same, so for either method the $1\frac{1}{2}$ hours must be changed to 90 minutes.

Remember:
Numbers written in ratios must always be in the same units.

The ratio method

Let x be the number of bottles filled.

The ratio of bottles to time is:

$$500 : 15$$
$$x : 90$$

$$\frac{x}{90} = \frac{500}{15}$$

$$\text{so } x = \frac{500 \times 90}{15} = 3000$$

So 3000 bottles are filled in $1\frac{1}{2}$ hours.

The unitary method

In 15 minutes, 500 bottles are filled. Therefore, in 1 minute, $500 \div 15$ bottles are filled. This gives a number of $33\frac{1}{3}$ bottles per minute.

So, in 90 minutes, $90 \times 33\frac{1}{3}$ bottles are filled. Therefore, in $1\frac{1}{2}$ hours, 3000 bottles are filled.

Exercise 4.2

Use either the ratio method or the unitary method to solve the problems below.

1 A machine prints four books in 10 minutes. How many will it print in 2 hours?

2 A farmer plants five apple trees in 25 minutes. If he continues to work at a constant rate, how long will it take him to plant 200 trees?

3 A television set uses 3 units of electricity in 2 hours. How many units will it use in 7 hours? Give your answer to the nearest unit.

4 A bricklayer lays 1500 bricks in an 8-hour day. Assuming he continues to work at the same rate, calculate:

a how many bricks he would expect to lay in a five-day week,

b how long, to the nearest hour, it would take him to lay 10 000 bricks.

5 A machine used to paint white lines on a road uses 250 litres of paint for each 8 km of road marked. Calculate:

a how many litres of paint would be needed for 200 km of road,

b what length of road could be marked with 4000 litres of paint.

If the information is given in the form of a ratio, the method of solution is the same.

Example   Tin and copper are mixed in the ratio 8:3. How much tin is needed to mix with 36 g of copper?

The ratio method

Let x grams be the mass of tin needed.

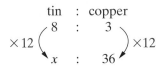

Remember:
Ratios stay the same if you multiply or divide the ratio by the same number

The metals must remain in the same ratio so we need to multiply each by the same number.

The amount of copper (36 g) is 12 times 3, so the amount of tin must also be multiplied by 12.

$$8 \times 12 = 96$$

Therefore 96 g of tin is needed.

The unitary method

3 g of copper mixes with 8 g of tin.
1 g of copper mixes with $\frac{8}{3}$ g of tin.
So 36 g of copper mixes with $36 \times \frac{8}{3}$ g of tin.
Therefore 36 g of copper mixes with 96 g of tin.

Exercise 4.3

Use either the ratio method or the unitary method to solve the problems below.

1 A production line produces 8 cars in 3 hours.
  a Calculate how many cars it will produce in 48 hours.
  b Calculate how long it will take to produce 1000 cars.
2 A machine produces six golf balls in fifteen seconds. Calculate how many are produced in:
  a 5 minutes   **b** 1 hour   **c** 1 day
3 A cassette recorder uses 0.75 units of electricity in 90 minutes. Calculate:
  a how many units it will use in 8 hours,
  b how long it will operate for 15 units of electricity.
4 A combine harvester takes 2 hours to harvest a 3 hectare field. If it works at a constant rate, calculate:
  a how many hectares it will harvest in 15 hours,
  b how long it will take to harvest a 54 hectare field.
5 A road-surfacing machine can re-surface 8 m of road in 40 seconds. Calculate how long it will take to re-surface 18 km of road, at the same rate.

Exercise 4.4

1 Sand and gravel are mixed in the ratio 5:3 to form ballast.
  a How much gravel is mixed with 750 kg of sand?
  b How much sand is mixed with 750 kg of gravel?
2 A recipe uses 150 g butter, 500 g flour, 50 g sugar and 100 g currants to make 18 small cakes.
  a How much of each ingredient will be needed to make 6 dozen cakes?
  b How many whole cakes could be made with 1 kg of butter?

3 A paint mix uses red and white paint in a ratio of 1:12.
  a How much white paint will be needed to mix with 1.4 litres of red paint?
  b If a total of 15.5 litres of paint is mixed, calculate the amount of white paint and the amount of red paint used. Give your answers to the nearest 0.1 litre.
4 A tulip farmer sells sacks of mixed bulbs to local people. The bulbs develop into two different colours of tulips, red and yellow. The colours are packaged in a ratio of 8:5 respectively.
  a If a sack contains 200 red bulbs, calculate the number of yellow bulbs.
  b If a sack contains 351 bulbs in total, how many of each colour would you expect to find?
  c One sack is packaged with a bulb mixture in the ratio 7:5 by mistake. If the sack contains 624 bulbs, how many more yellow bulbs would you expect to have, compared with a normal sack of 624 bulbs?
5 A pure fruit juice is made by mixing the juices of oranges and mangoes in the ratio of 9:2.
  a If 189 litres of orange juice are used, calculate the number of litres of mango juice needed.
  b If 605 litres of the juice are made, calculate the number of litres of orange juice and mango juice used.

Dividing a quantity in a given ratio

Examples Divide 20 m in the ratio 3:2.

The ratio method
3:2 gives 5 parts.

$$\frac{3}{5} \times 20\,m = 12\,m$$
$$\frac{2}{5} \times 20\,m = 8\,m$$

Therefore 20 m divided in the ratio 3:2 is 12 m:8 m.

The unitary method
3:2 gives 5 parts.
5 parts are equivalent to 20 m.
1 part is equivalent to $\frac{20}{5}$ m (i.e. 4 m).
Therefore 3 parts are equivalent to 3×4 m, that is 12 m; and 2 parts are equivalent to 2×4 m, that is 8 m.

A factory produces cars in red, blue, white and green in the ratio 7:5:3:1. Out of a production of 48 000 cars how many are white?

$7 + 5 + 3 + 1$ gives a total of 16 parts.

Therefore the total number of white cars $= \dfrac{48\,000}{16} \times 3 = 9000$.

Exercise 4.5

1 Divide 150 in the ratio 2:3.
2 Divide 72 in the ratio 2:3:4.
3 Divide 5 kg in the ratio 13:7.
4 Divide 45 minutes in the ratio 2:3.
5 Divide 1 hour in the ratio 1:5.
6 $\frac{7}{8}$ of a can of coke is water, the rest is syrup. What is the ratio of water to syrup?
7 $\frac{5}{9}$ of a litre carton of orange is pure orange juice, the rest is water. How many millilitres of each are in the carton?

8 55% of students in a school are boys.
  a What is the ratio of boys to girls?
  b How many boys and how many girls are there if the school has 800 students?
9 A piece of wood is cut in the ratio of 2:3. What fraction of the length is the longer piece?
10 If the original piece of wood in question 9 is 80 cm long, how long is the shorter piece?

SUMMARY

By the time you have finished this chapter you should know:

■ how to identify and generate equivalent fractions, for example $\frac{4}{6}$, $\frac{8}{12}$, $\frac{12}{18}$, $\frac{20}{30}$ and $\frac{28}{42}$ are all equivalent fractions, since by dividing numerator and denominator by the same factor, they each cancel to $\frac{2}{3}$.

■ that **ratios** can be simplified in the same way as fractions, for example,

6:18 is the same as 1:3   (divide by 6)
14:35 is the same as 2:5   (divide by 7)

■ that problems involving **direct proportion** can be solved either by the **ratio method** or by the **unitary method**, for example tin and copper are mixed in the ratio 7:3 – how much tin is needed to mix with 24 g of copper?

Ratio method: if x grams is the mass of tin needed, $\dfrac{x}{24} = \dfrac{7}{3}$, so

$$x = \frac{7 \times 24}{3} = 56, \quad \text{i.e. } 56\,\text{g of tin is needed}$$

Unitary method: 3 g of copper mixes with 7 g of tin, so
1 g of copper mixes with $\frac{7}{3}$ g of tin.
Therefore 24 g of copper mixes with $\frac{7}{3} \times 24$ g of tin, i.e. 56 g of tin.

■ how to divide a quantity in a given ratio, for example to divide 60 m in the ratio 3:2 by the unitary method,

3:2 gives 5 parts. 5 parts are equivalent to 60 m, so 1 part is equivalent to $60 \div 5$ m, i.e. 12 m.
Therefore 3 parts are 3×12 m, i.e. 36 m, and 2 parts are 2×12, i.e. 24 m.

Exercise 4A

1 A ruler 30 cm long is broken into two parts in the ratio 8:7. How long are the two parts?
2 A recipe needs 400 g of flour to make 8 cakes. How much flour would be needed to make two dozen cakes?
3 To make 6 jam tarts, 120 g of jam is needed. How much jam is needed to make 10 tarts?
4 The scale of a map is 1:25 000.
  a Two villages are 8 cm apart on the map. How far apart are they in real life? Give your answer in kilometres.
  b The distance from a village to the edge of a lake is 12 km in real life. How far apart would they be on the map? Give your answer in centimetres.
5 A motorbike uses a petrol and oil mixture in the ratio 13:2.
  a How much of each is there in 30 litres of mixture?
  b How much petrol would be mixed with 500 ml of oil?

6 a A model car is a $\frac{1}{40}$ scale model. Express this as a ratio.
  b If the length of the real car is 5.5 m, what is the length of the model car?
7 An aunt gives a brother and sister £2000 to be divided in the ratio of their ages. If the girl is 13 years old and the boy 12 years old, how much will each get?
8 The angles of a triangle are in the ratio $2:5:8$. Find the size of each of the angles.

Exercise 4B

1 A piece of wood is cut in the ratio $3:7$.
  a What fraction of the whole is the longer piece?
  b If the wood is 1.5 m long, how long is the shorter piece?
2 A recipe for two people requires $\frac{1}{4}$ kg of rice to 150 g of meat.
  a How much meat would be needed for five people?
  b How much rice would there be in 1 kg of the final dish?
3 The scale of a map is $1:10\,000$.
  a Two rivers are 4.5 cm apart on the map. How far apart are they in real life? Give your answer in metres.
  b Two towns are 8 km apart in real life. How far apart are they on the map? Give your answer in centimetres.
4 a A model train is a $\frac{1}{25}$ scale model. Express this as a ratio.
  b If the length of the model engine is 7 cm, what is the true length of the engine?
5 Divide 3 tonnes in the ratio $2:5:13$.
6 The ratio of the angles of a quadrilateral is $2:3:3:4$. Calculate the size of each of the angles.
7 The ratio of the interior angles of a pentagon is $2:3:4:4:5$. Calculate the size of the largest angle.
8 The lengths of the sides of a rectangle are in the ratio $2:1$. If the area of the rectangle is 98 cm^2, calculate the length and width of the rectangle.

Exercise 4C

You will need:
• squared paper

The two rectangles shown below are mathematically similar. The dimensions of rectangle A are twice those of rectangle B.

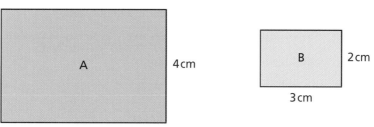

1 The ratio of length A to length B is $2:1$. What is the ratio of area A to area B?
2 Draw three more pairs of rectangles where the dimensions of A are twice those of B. What is the ratio of their areas?
3 Draw three pairs of rectangles where the dimensions of A are three times those of B. What is the ratio of their areas?
4 What is the ratio of areas if the dimensions are $4:1$?
5 What is the ratio of areas if the dimensions are $n:1$?
  Investigate this thoroughly, showing your working and reasoning clearly.

Exercise 4D

Leonardo da Vinci was a great Italian Renaissance artist who lived from 1452 to 1519. One of his famous sketches is shown on the right.

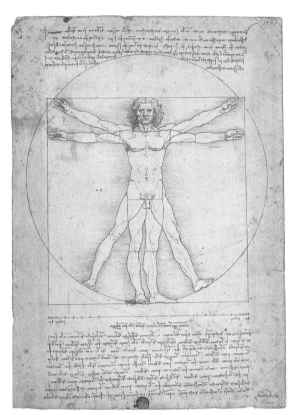

You will need:
- tape measure
- computer with a spreadsheet package installed
- ruler

Leonardo's most famous works include the *Mona Lisa* and *The Last Supper*. Although he was best known for his art, he was also an accomplished architect, engineer and designer. His sketch books show ideas for dozens of inventions, many of them way ahead of their time. These include designs for bridges, parachutes and helicopters.

The sketch implies that our bodies have a certain proportion.

1 For all the pupils in your class, collect the following data:
   a their height,
   b their arm span measured from the tips of the middle fingers.
2 Enter the results into a spreadsheet and plot a scatter graph of height against arm span.
3 What can you deduce from the results of the graph?
4 Draw a line of best fit through the graph (either by hand or using the spreadsheet itself, if possible).
5 Calculate the gradient of the line of best fit.
6 What does this gradient tell you about the ratio of height : arm span?

Exercise 4E

Use a library or the internet to find out more about Zhoubi Zuanjing's book, *Nine Chapters of the Mathematician's Art*, with special reference to chapters 2, 3 and 6, which are concerned with ratio and proportion.

5 Use of a calculator

You should try to make full use of your calculator's potential. Nowadays basic calculators, scientific calculators and graphical calculators are all available; they have a long history.

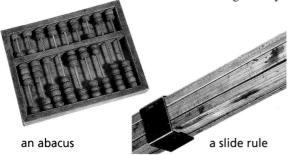

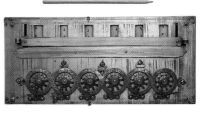

an abacus   a slide rule   Blaise Pascal's calculator   an early calculator

Calculators are of no help to you if you don't make use of their potential. This chapter is aimed at familiarising you with some of the basic **operations**.

The four basic operations

Examples   **Note.** Always make an estimate before pressing any keys.

Using a calculator, work out the answer to $12.3 + 14.9$.

$12.3 + 14.9 =$ (Estimate 27)

[1] [2] [.] [3] [+] [1] [4] [.] [9] [=] 27.2

Using a calculator, work out the answer to 16.3×10.8.

$16.3 \times 10.8 =$ (Estimate 160)

[1] [6] [.] [3] [×] [1] [0] [.] [8] [=] 176.04

Using a calculator, work out the answer to $24.1 \div -3.3$.

$24.1 \div -3.3 =$ (Estimate -8)

[2] [4] [.] [1] [÷] [3] [.] [3] [±] [=] -7.30

Using a calculator, work out the answer to $-15.8 - 7.4$.

$-15.8 - 7.4 =$ (Estimate -23)

[1] [5] [.] [8] [±] [−] [7] [.] [4] [=] -23.2

Exercise 5.1

Using a calculator, work out the answers to the following.

1 a $9.7 + 15.3$ **b** $13.6 + 9.08$ **c** $12.9 + 4.92$ **d** $115.0 + 6.24$ **e** $86.13 + 48.2$ **f** $108.9 + 47.2$

2 a $15.2 - 2.9$ **b** $12.4 - 0.5$ **c** $19.06 - 20.3$ **d** $4.32 - 4.33$ **e** $-9.1 - 21.2$ **f** $-6.3 - 2.1$
  g $-28 - (-15)$ **h** $-2.41 - (-2.41)$

3 a 9.2×8.7 **b** 14.6×8.1 **c** $4.1 \times 3.7 \times 6$ **d** $9.3 \div 3.1$ **e** $14.2 \times (-3)$ **f** $15.5 \div (-5)$
  g $-2.2 \times (-2.2)$ **h** $-20 \div (-4.5)$

The order of operations

When carrying out calculations, care must be taken to ensure that they are carried out in the correct order.

Examples Use a scientific calculator to work out the answer to $2 + 3 \times 4$.

$2 + 3 \times 4 =$ 2 + 3 × 4 = 14

Use a scientific calculator to work out the answer to $(2 + 3) \times 4$.

$(2 + 3) \times 4 =$ (2 + 3) × 4 = 20

The reason why different answers are obtained is because, in mathematics, different operations have different **priorities**. The order is as follows:

1 brackets
2 powers
3 multiplication/division
4 addition/subtraction.

In the first example 3×4 is evaluated (worked out) first and then added to the 2, whilst in the second example $(2 + 3)$ is evaluated first, and then multiplied by 4.

Exercise 5.2

In each part of questions 1–3, evaluate the answer:

i) in your head

ii) using a scientific calculator.

1 a $8 \times 3 + 2$ **b** $4 \div 2 + 8$ **c** $6 - 3 \times 4$ **d** $10 - 6 \div 3$
2 a $7 \times 2 + 3 \times 2$ **b** $9 + 3 \times 8 - 1$ **c** $36 - 9 \div 3 - 2$ **d** $4 + 3 \times 7 - 6 \div 3$
3 a $(4 + 5) \times 3$ **b** $8 \times (12 - 4)$ **c** $3 \times (8 + 3) - 3$ **d** $24 \div 3 \div (10 - 5)$

In questions 4 and 5:
i) copy the calculation and put in any brackets which are needed to make it correct,
ii) check your answers using a scientific calculator.

4 a $6 \times 2 + 1 = 18$ **b** $1 + 3 \times 5 = 16$ **c** $8 + 6 \div 2 = 7$ **d** $3 + 2 \times 4 - 1 = 15$
5 a $12 \div 4 - 2 + 6 = 7$ **b** $12 \div 4 - 2 + 6 = 12$ **c** $12 \div 4 - 2 + 6 = -5$ **d** $12 \div 4 - 2 + 6 = 1.5$

Powers

Calculators, particularly the more advanced ones, have a large number of keys and **functions**. Many of them are simply more efficient ways of doing calculations.

Examples Using a calculator, evaluate 7^2.

7^2 means 7 to the power of 2, or 7 squared, and can be written as 7×7. The x^2 key is the 'squared' key, so 7^2 can be entered into the calculator as follows:

7 x^2 = 49

Using a calculator, evaluate 3^5.

> Your calculator may have an x^y key instead. It does exactly the same job.

3^5 means 3 to the power of 5 and can be written as $3 \times 3 \times 3 \times 3 \times 3$ which is 243. But this is not an efficient way of doing the calculation on the calculator.

The y^x key is the 'to the power of' key.

So 3^5 can be entered into the calculator as follows:

| 3 | | y^x | | 5 | | = | 243

Evaluate the following using a scientific calculator:

$$\frac{90 + 38}{4^3} = \boxed{(} \boxed{9} \boxed{0} \boxed{+} \boxed{3} \boxed{8} \boxed{)} \boxed{\div} \boxed{4} \boxed{y^x} \boxed{3} \boxed{=} 2$$

Exercise 5.3

Using a scientific calculator, evaluate the following.

1 **a** $\dfrac{9 + 3}{6}$ **b** $\dfrac{30 - 6}{5 + 3}$ **c** $\dfrac{15 \times 2}{7 + 8} + 2$ **d** $\dfrac{7 + 2 \times 4}{7 - 2} - 3$

2 **a** $\dfrac{4^2 - 6}{2 + 8}$ **b** $\dfrac{3^2 + 4^2}{5}$ **c** $\dfrac{3^3 \times 4^4}{12^2} + 2$ **d** $\dfrac{(6 + 3) \times 4}{2^3} - 2 \times 3$

SUMMARY

By the time you have completed this chapter you should know:

■ how to use the following calculator buttons: $\boxed{+}$ $\boxed{-}$ $\boxed{\times}$ $\boxed{\div}$ $\boxed{\pm}$ $\boxed{x^y}$ $\boxed{x^2}$ (or the appropriate keys that perform these functions on your calculator – for example, the $\boxed{x^y}$ key may be the $\boxed{y^x}$ key)

■ which **operations** take priority when doing a calculation:

- brackets
- powers
- multiplication/division
- addition/subtraction

Exercise 5A

1 Work out the answers to the following.
   a $-5.1 + 10$ **b** $4.8 - (-8.2)$ **c** $3.3 \times (-4.2)$ **d** $-15 \div (-4)$

2 Evaluate the following.
   a $3 + 5 \times 2$ **b** $3 \times 3 + 4 \times 4$ **c** $3 + 3 \times 4 + 4$ **d** $18 \div 2 \div (5 - 2)$

3 Copy the following, putting in brackets if necessary to make the statement correct.
   a $7 - 4 \times 2 = 6$ **b** $12 + 3 \times 3 + 4 = 33$ **c** $5 + 5 \times 6 - 4 = 20$ **d** $5 + 5 \times 6 - 4 = 56$

4 Evaluate the following.

   a $\dfrac{2^4 - 3^2}{2}$ **b** $\dfrac{(8 - 3) \times 3}{5} + 7$

Exercise 5B

1 Work out the answers to the following.

  a $-6.1 + 4$ **b** $4.2 - (-5.2)$ **c** -3.6×4.1 **d** $-18 \div (-2.5)$

2 Evaluate the following.

  a $12 + 6 \div 2$ **b** $3 + 4 \div 2 \times 4$ **c** $6 + 3 \times 4 - 5$ **d** $14 \times 2 \div (9 - 2)$

3 Copy the following, putting in brackets if necessary to make the statement correct.

  a $7 - 5 \times 3 = 6$ **b** $16 + 4 \times 2 + 4 = 40$ **c** $4 + 5 \times 6 - 1 = 45$ **d** $1 + 5 \times 6 - 6 = 30$

4 Evaluate the following.

  a $\dfrac{3^3 - 4^2}{2}$ **b** $\dfrac{(15 - 3) \div 3}{2} + 7$

Exercise 5C

Using four 4s each time (and no other number), write down calculations that will give each of the numbers from 1 to 30 as the answer. You may use any mathematical operation that you know.

  For example, to give an answer of 8:

$$(4 + 4) \times 4 \div 4 = 8$$
$$\sqrt{4} + \sqrt{4} + \sqrt{4} + \sqrt{4} = 8$$

Check that you have written each of your calculations correctly by entering them into a calculator.

Exercise 5D

Use your calculator to find the fallacy in the following sequence of equations.

$$25 - 45 = 16 - 36$$
$$25 - 45 + \tfrac{81}{4} = 16 - 36 + \tfrac{81}{4}$$
$$25 - 45 + \left(\tfrac{9}{2}\right)^2 = 16 - 36 + \left(\tfrac{9}{2}\right)^2$$
$$\left(5 - \tfrac{9}{2}\right)^2 = \left(4 - \tfrac{9}{2}\right)^2$$
$$5 - \tfrac{9}{2} = 4 - \tfrac{9}{2} \qquad \text{[square root]}$$
$$5 - \tfrac{9}{2} + \tfrac{9}{2} = 4 - \tfrac{9}{2} + \tfrac{9}{2}$$
$$5 = 4$$

Exercise 5E

Using the internet as a resource, produce a display which describes the history of the calculator and calculation machines.

OR

Investigate the use of the abacus as a calculator.

You will need:
• computer with internet access

6 Linear equations

Isaac Newton discovered the universal law of gravitation, which is written as:

$$F = G\frac{m_1 m_2}{d^2}$$

Albert Einstein found the connection between matter and energy and wrote it as:

$$E = mc^2$$

Andrew Wiles proved Fermat's last theorem:

$$x^n \neq y^n + z^n$$

where x, y and z are whole numbers and $n > 2$.

Charlie Chaplin once said of Albert Einstein and himself, 'I am famous because everyone understands me, he is famous because nobody does.'

Pierre de Fermat was a seventeenth-century mathematician who often wrote to other mathematicians, challenging them to find proofs for his theorems. He claimed that he didn't have the room in his workbook to write down the proof of his last theorem, and so the problem remained unsolved for hundreds of years – until Andrew Wiles came along.

The three equations above look simple but the mathematics leading to them is extremely complex.

This chapter considers only **linear equations**, which are easier than those mentioned above. A linear equation is one which, if drawn on a graph, would give a straight line.

Revision

An equation represents two quantities which are equal to each other. To help visualise what an equation is and how it can be manipulated, it can be thought of as a pair of scales.

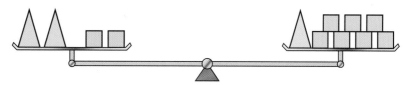

In the scales above there are two different types of object: △ and ☐.

The left-hand side of the scales balances the right-hand side, i.e. the masses on both sides are equal.

If a 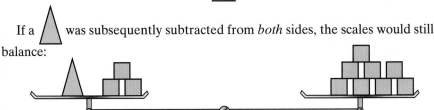 was added to *both* sides, the scales would still balance:

If a was subsequently subtracted from *both* sides, the scales would still balance:

If *both* sides were doubled, the scales would still balance:

In fact the scales would always balance as long as what was done to one side was also done to the other. By knowing this it is possible to solve equations, that is, to work out the value of the unknown quantity.

Example

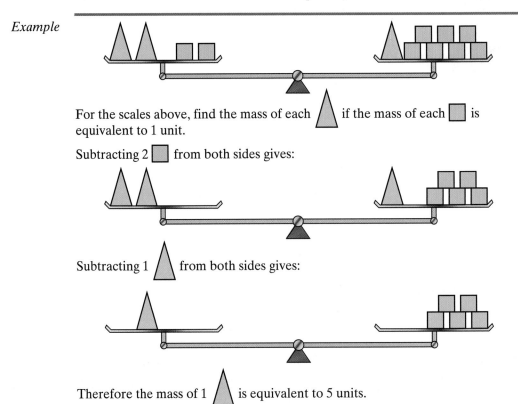

For the scales above, find the mass of each if the mass of each is equivalent to 1 unit.

Subtracting 2 from both sides gives:

Subtracting 1 from both sides gives:

Therefore the mass of 1 is equivalent to 5 units.

However, drawing scales each time can be a laborious process. Algebra is therefore used instead of diagrams. The problem above can be written as

$2x + 2 = x + 7$ where x represents the mass of 1 △. Therefore solving the equation gives:

$$2x + 2 = x + 7$$
$$2x = x + 5$$
$$x = 5$$

Example Solve the following linear equations.
a $3x + 8 = 14$
b $12 = 20 + 2x$
c $3(p + 4) = 21$
d $4(x - 5) = 7(2x - 5)$

a $3x + 8 = 14$
$$3x = 14 - 8$$
$$3x = 6$$
$$x = 6 \div 3$$
$$x = 2$$
b $12 = 20 + 2x$
$$12 - 20 = 2x$$
$$-8 = 2x$$
$$-8 \div 2 = x$$
$$-4 = x$$
c $3(p + 4) = 21$
$$3p + 12 = 21$$
$$3p = 21 - 12$$
$$3p = 9$$
$$p = 9 \div 3$$
$$p = 3$$
d $4(x - 5) = 7(2x - 5)$
$$4x - 20 = 14x - 35$$
$$4x - 20 + 35 = 14x$$
$$4x + 15 = 14x$$
$$15 = 14x - 4x$$
$$15 = 10x$$
$$15 \div 10 = x$$
$$1.5 = x$$

Exercise 6.1

Solve the following linear equations.

1 a $5a - 2 = 18$ **b** $7b + 3 = 17$ **c** $9c - 12 = 60$ **d** $6d + 8 = 56$
e $4e - 7 = 33$ **f** $12f + 4 = 76$
2 a $4a = 3a + 7$ **b** $8b = 7b - 9$ **c** $7c + 5 = 8c$ **d** $5d - 8 = 6d$
3 a $3a - 4 = 2a + 7$ **b** $5b + 3 = 4b - 9$ **c** $8c - 9 = 7c + 4$ **d** $3d - 7 = 2d - 4$
4 a $6a - 3 = 4a + 7$ **b** $5b - 9 = 2b + 6$ **c** $7c - 8 = 3c + 4$ **d** $11d - 10 = 6d - 15$
5 a $3(a + 1) = 9$ **b** $5(b - 2) = 25$ **c** $8 = 2(c - 3)$ **d** $14 = 4(3 - d)$
e $21 = 3(5 - e)$ **f** $36 = 9(5 - 2f)$

Exercise 6.2

Solve the following linear equations.

1 a $3x = 2x - 4$
  d $p - 8 = 3p$
b $5y = 3y + 10$
  e $3y - 8 = 2y$
c $2y - 5 = 3y$
  f $7x + 11 = 5x$

2 a $3x - 9 = 4$
  d $4y + 5 = 3y - 3$
b $4 = 3x - 11$
  e $8y - 31 = 13 - 3y$
c $6x - 15 = 3x + 3$
  f $4m + 2 = 5m - 8$

3 a $7m - 1 = 5m + 1$
  d $6x + 9 = 3x - 54$
b $5p - 3 = 3 + 3p$
  e $8 - 3x = 18 - 8x$
c $12 - 2k = 16 + 2k$
  f $2 - y = y - 4$

4 a $-2(x - 5) = 6$
  d $2(x + 1) = 3(x - 5)$
  g $7(7 + 2x) = 3(9x - 1)$
b $3(x - 2) = 4(4x - 8)$
  e $5(x - 4) = 3(x + 2)$
  h $6(2x + 3) = 4(4x - 2)$
c $18 = -2(y - 1)$
  f $-4(3 + y) = 2(y + 1)$

> **Remember:**
> *When you expand an expression, you are getting rid of the brackets. When you factorise an expression, you are putting in brackets.*

Factorising

The use of brackets has been shown above. If an expression such as $3(x + 2)$ is multiplied out to become $3x + 6$, this process is called **expansion**. The opposite to expanding is known as **factorising**. For example

$$3(x + 2) \rightarrow 3x + 6 \text{ is expansion}$$
$$3x + 6 \rightarrow 3(x + 2) \text{ is factorising}$$

To factorise we need to find the highest common factor to all the terms. This is then written outside the brackets.

Examples   Factorise $4x + 6$.

The highest common factor of $4x$ and 6 is 2.
Therefore $4x + 6 = 2(2x + 3)$.

....

Factorise $5a + 2ab + a$.

Simplifying the expression gives $6a + 2ab$.
The highest common factor of $6a$ and $2ab$ is $2a$.
Therefore $6a + 2ab = 2a(3 + b)$.

Factorising can also be used to simplify equations before solving them.

Examples   Solve the equation $5x + 10 = 25$.

Factorise: $5(x + 2) = 25$
Divide by 5: $5(x + 2) \div 5 = 25 \div 5$
                  $x + 2 = 5$
Subtract 2: $x + 2 - 2 = 5 - 2$
                    $x = 3$

....

Solve the equation $3 = 6x - 18$.

Factorise: $3 = 6(x - 3)$
Divide by 6: $3 \div 6 = 6(x - 3) \div 6$
               $\frac{1}{2} = x - 3$
Add 3: $\frac{1}{2} + 3 = x - 3 + 3$
            $3\frac{1}{2} = x$

Exercise 6.3

1 Factorise the following.

   a $3a + 12$ **b** $5x + 15$ **c** $27 + 12y$

   d $4a + 8x + 6$ **e** $9p - 6q$ **f** $8g + 12h - 16k$

   g $7a - 21 + 14y$ **h** $42x - 28y$ **i** $36a + 24b$

   j $32x - 48y + 16$

2 Factorise the following.

   a $6ab + 4a$ **b** $7xy + 28y$ **c** $15mn + 20m$

   d $9pq - 27q$ **e** $t - 6tu$ **f** $18a - 24ab$

   g $14x + 35xy$ **h** $16pr - 36pq$

3 By factorising, solve the following equations.

   a $2x + 4 = 10$ **b** $36 + 6y = 18$ **c** $15p + 20 = 50$

   d $7x - 21 = 14$ **e** $-9a + 12 = 3$ **f** $-6p + 14 = -4$

   g $-15a - 20 = 10$ **h** $-14x + 42 = 0$

Constructing equations

In many cases, when dealing with the practical applications of mathematics, equations need to be constructed before they can be solved. Often the information is either given within the problem itself or in a diagram.

Examples   Find the size of each of the angles in the triangle below by constructing an equation and solving it to find the value of x.

> The diagrams in this chapter are not drawn to scale.

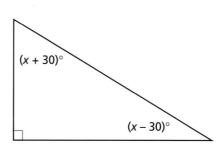

The sum of the angles of a triangle equals $180°$.

$$x + 30 + x - 30 + 90 = 180$$
$$2x + 90 = 180$$
$$2x = 90$$
$$x = 45$$

The three angles are therefore: $90°$
$$(x + 30)° = 75°$$
$$(x - 30)° = 15°$$

Check: $90° + 75° + 15° = 180°$.

Construct an equation and solve it to find the value of *x*.

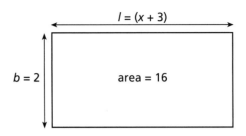

The formula used to calculate the area of a rectangle is $A = lb$. By substituting $l = (x + 3)$, $b = 2$ and $A = 16$ in the formula we have:

$$A = lb$$
$$16 = (x + 3) \times 2$$
$$16 = 2x + 6$$
$$16 - 6 = 2x$$
$$10 = 2x$$
$$5 = x$$

(so $l = 8$)

Exercise 6.4

In questions 1–3:

- construct an equation in terms of *x*,
- solve the equation,
- calculate the value of each of the angles.

Remember to check your answers.

1 a

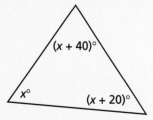

b

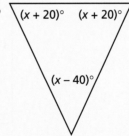

c

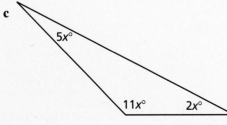

d

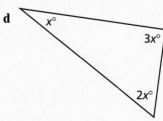

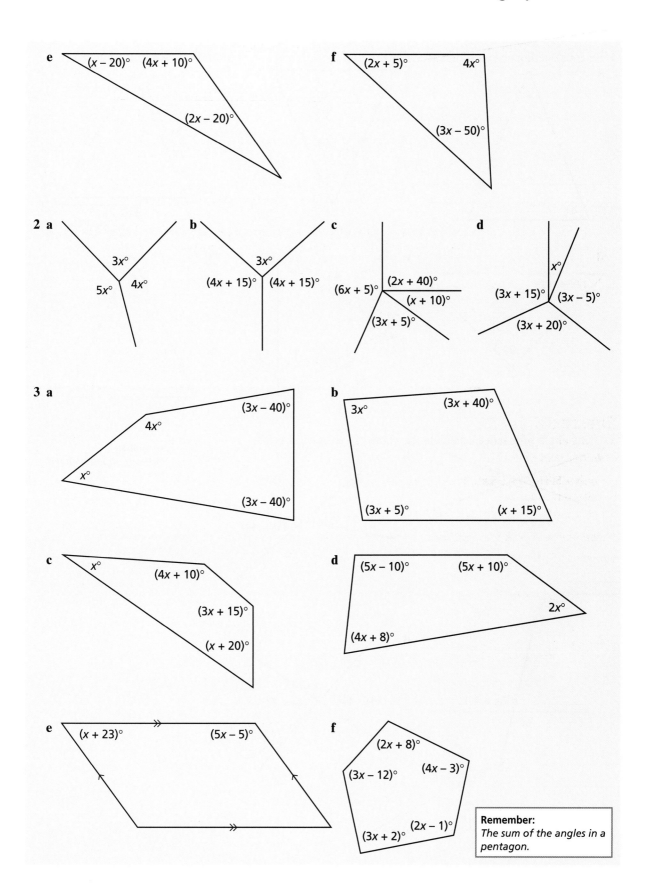

e $(x - 20)°$ $(4x + 10)°$ $(2x - 20)°$

f $(2x + 5)°$ $4x°$ $(3x - 50)°$

2 a $3x°$ $5x°$ $4x°$

b $3x°$ $(4x + 15)°$ $(4x + 15)°$

c $(6x + 5)°$ $(2x + 40)°$ $(x + 10)°$ $(3x + 5)°$

d $x°$ $(3x + 15)°$ $(3x - 5)°$ $(3x + 20)°$

3 a $(3x - 40)°$ $4x°$ $x°$ $(3x - 40)°$

b $3x°$ $(3x + 40)°$ $(3x + 5)°$ $(x + 15)°$

c $x°$ $(4x + 10)°$ $(3x + 15)°$ $(x + 20)°$

d $(5x - 10)°$ $(5x + 10)°$ $2x°$ $(4x + 8)°$

e $(x + 23)°$ $(5x - 5)°$

f $(2x + 8)°$ $(3x - 12)°$ $(4x - 3)°$ $(3x + 2)°$ $(2x - 1)°$

Remember:
The sum of the angles in a pentagon.

4 By constructing an equation and solving it, find the value of x in each of these isosceles triangles.

a

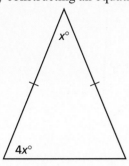

$x°$

$4x°$

b

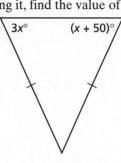

$3x°$ $(x + 50)°$

c

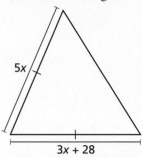

$5x$

$3x + 28$

d

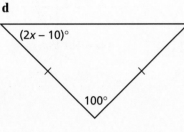

$(2x - 10)°$

$100°$

e

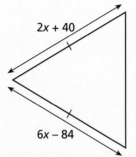

$2x + 40$

$6x - 84$

f

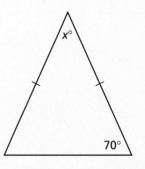

$x°$

$70°$

5 Using angle properties, calculate the value of x in each of these.

a

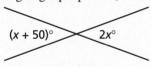

$(x + 50)°$ $2x°$

b

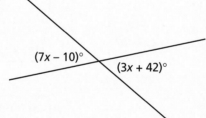

$(7x - 10)°$

$(3x + 42)°$

> **Remember:**
> *Vertically opposite angles.*

c

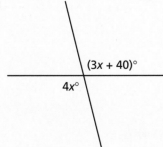

$(3x + 40)°$

$4x°$

d

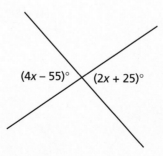

$(4x - 55)°$ $(2x + 25)°$

6 Calculate the value of x in each of these.

a

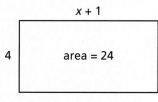

$x + 1$

4 | area = 24

b

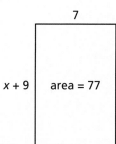

7

$x + 9$ | area = 77

c

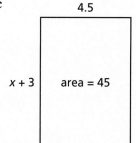

4.5

$x + 3$ | area = 45

d

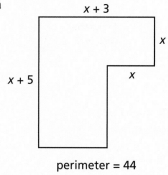

$x + 0.4$

area = 5.7 | 3.8

e

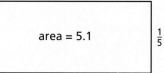

$x + \frac{1}{2}$

area = 5.1 | $\frac{1}{5}$ | x

f

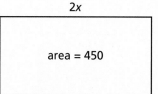

$2x$

x | area = 450

7 Calculate the value of x in each of these.

a

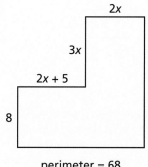

$x + 3$

x

x

$x + 5$

perimeter = 44

b

$2x$

$3x$

$2x + 5$

8

perimeter = 68

c

x

x

$3x$

$5x$

perimeter = 108

d

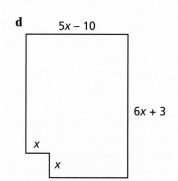

$5x - 10$

$6x + 3$

x

x

perimeter = 140

e

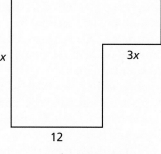

$3x$

$7x$

$3x$

12

perimeter = 224

f

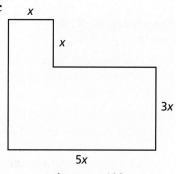

$2x$

60°

60°

perimeter = 150

Examples A number is doubled, then 5 is subtracted and the result is 17. Find the number.

Let x be the unknown number.

$$2x - 5 = 17$$
$$2x = 17 + 5$$
$$2x = 22$$
$$x = 11$$

The number is 11.

....

3 is added to a number. The result is multiplied by 8. If the answer is 64, calculate the value of the original number.

Let x be the unknown number.

$$8(x + 3) = 64 \qquad \text{or} \qquad 8(x + 3) = 64$$
$$8x + 24 = 64 \qquad\qquad\qquad x + 3 = 64 \div 8$$
$$8x = 64 - 24 \qquad\qquad\qquad x + 3 = 8$$
$$8x = 40 \qquad\qquad\qquad\qquad x = 8 - 3$$
$$x = 40 \div 8 \qquad\qquad\qquad\qquad x = 5$$
$$x = 5$$

The original number is 5.

Exercise 6.5

In the following questions:
i) construct an equation in terms of x,
ii) solve the equation.

1 a A number is trebled and then 7 is added to it. If the total is 28, find the number.
  b Multiply a number by 4 and then add 5 to it. If the total is 29, find the number.
  c If 31 is the result of adding 1 to 5 times a number, find the number.
  d Double a number and then subtract 9. If the answer is 11, what is the number?
  e If 9 is the result of subtracting 12 from 7 times a number, find the number.
2 a Add 3 to a number and then double the result. If the total is 22, find the number.
  b 27 is the answer when you add 4 to a number and then treble it. What is the number?
  c Subtract 1 from a number and multiply the result by 5. If the answer is 35, what is the number?
  d Add 3 to a number. If the result of multiplying this total by 7 is 63, find the number.
  e Add 3 to a number. Quadruple the result. If the answer is 36, what is the number?
3 a Gabriella is x years old. Her brother is 8 years older than her and her sister is 12 years younger than her. If their total age is 50 years, how old are they?
  b A series of mathematics textbooks consists of four volumes. The first volume has x pages, the second 54 pages more. The third and fourth volume each have 32 pages more than the second. If the total number of pages in all four volumes is 866, calculate the number of pages in each of the volumes.
  c The five interior angles of a pentagon (in degrees) are x, $x + 30$, $2x$, $2x + 40$ and $3x + 20$. If the sum of the interior angles of a pentagon is 540°, calculate the size of each of the angles.
  d A hexagon consists of three interior angles of equal size and a further three which are double the size. If the sum of all six angles is 720°, calculate the size of each of the angles.
  e Four of the exterior angles of an octagon are the same size. The other four are twice as big. If the sum of the exterior angles is 360°, calculate the size of the *interior angles*.

SUMMARY

By the time you have completed this chapter you should know:

- what is meant by a linear equation, for example $x + 3 = 7$, $5 = 3 - p$ and $6m - 2 = 40$ are linear equations; they can be plotted on a graph and give a straight line (hence 'linear'). When solved they have only one solution
- how to solve a **linear equation**, for example

$$x + 3 = 7 \qquad\qquad 5 = 3 - p \qquad 6m - 2 = 40$$
$$x = 7 - 3 \qquad 5 + p = 3 \qquad\qquad 6m = 40 + 2$$
$$x = 4 \qquad\qquad p = 3 - 5 \qquad\qquad 6m = 42$$
$$p = -2 \qquad\qquad m = 42 \div 6$$
$$m = 7$$

- how to solve linear equations involving negatives and brackets, for example

$$3(a - 5) = 9 \qquad\qquad -5(x - 3) = -3(x + 2)$$
$$3a - 15 = 9 \qquad\qquad -5x + 15 = -3x - 6$$
$$3a = 9 + 15 \qquad\qquad 15 + 6 = -3x + 5x$$
$$3a = 24 \qquad\qquad 21 = 2x$$
$$a = 24 \div 3 \qquad\qquad 21 \div 2 = x$$
$$a = 8 \qquad\qquad 10.5 = x$$

- how to construct a linear equation, for example a total of 28 is obtained by adding 3 to a number and then doubling the result. Find the number.

  Let n represent the unknown number.
$$28 = 2(3 + n)$$
$$28 = 6 + 2n$$
$$28 - 6 = 2n$$
$$22 = 2n$$
$$11 = n$$
  Therefore the number is 11.

Exercise 6A

Solve the following equations.

1 a $x + 7 = 16$ **b** $2x - 9 = 13$
   c $8 - 4x = 24$ **d** $5 - 3x = -13$
2 a $7 - m = 4 + m$ **b** $5m - 3 = 3m + 11$
   c $6m - 1 = 9m - 13$ **d** $18 - 3p = 6 + p$
3 a $2(x - 4) = 24$ **b** $4(x - 3) = 7(x + 2)$
   c $28 = 2(3x + 8)$ **d** $24(x - 1) = 20(2x - 4)$
4 a $-3(x + 1) = 3$ **b** $-(4 - x) = 2(x - 2.5)$
   c $2(x - 4) = -(2x - 8)$ **d** $-2(x - 1) = -x + 3$

Exercise 6B

Solve the following equations.

1 a $y + 9 = 3$ **b** $3x - 5 = 13$
   c $12 - 5p = -8$ **d** $2.5y + 1.5 = 7.5$
2 a $5 - p = 4 + p$ **b** $8m - 9 = 5m + 3$
   c $11p - 4 = 9p + 15$ **d** $27 - 5r = r - 3$

3 a $2(t-1) = 15$ **b** $5(3-m) = 4(m-6)$
  c $15 = 2(x-1)$ **d** $16(t-2) = 5(2t+8)$

4 Find the size of each of the angles in the quadrilateral below by constructing an equation and solving it to find the value of x.

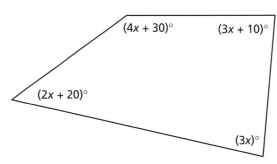

Exercise 6C

You will need:
• graphical calculator

Many graphical calculators have an *equation solver capability*.
Use a graphical calculator to check your solutions to exercise 6.1.
   For example, use a graphical calculator to solve the equation $2(3x-2) = 5$.
   The equation will probably need to be rearranged so that it is equal to zero:

$$2(3x-2) - 5 = 0$$

Enter the equation into the calculator (the example shows the screen of a T1-83).

The screen is explained as follows: solve$(2(3x-2) - 5, x, 0)$.

                    equation   variable   approximate
                                           solution

Press enter to give the solution:

Exercise 6D

Dops and dips

If the outside dots in the diagram above are joined they form a rectangle. The rectangle has an area of 6 square units (if the horizontal and vertical distance between the dots is said to be 1 unit).

If the rectangle is drawn then the diagram above has ten dops – **d**ots **o**n the **p**erimeter.

It also has two dips – **d**ots **i**nside the **p**erimeter.

1 Investigate the relationship between area and dops in rectangles with no dips.
2 Investigate the relationship between area, dips and dops in rectangles with a single line of dips.
3 Investigate other rectangles.

Exercise 6E

Fermat's Last Theorem by Simon Singh (published by Fourth Estate, London), the book which describes the work of many mathematicians that led to Professor Andrew Wiles' solution of a two hundred year old problem, is well worth reading. The BBC *Horizon* programme of the same name is outstanding – see it if you can.

> Fermat's last theorem states that $x^n + y^n = z^n$ has no whole number solutions except when $n = 1$ or 2.

7 Formulae and substitution

The formula $E = mc^2$ is one of the most recognisable formulae of all time. It was stated by an Austrian mathematician who spent some of his working life as a clerk in a patent office. However, his work in mathematics was to make him a major figure of the twentieth century.

That man was Albert Einstein. In fact, Einstein's equation is very simple. $E = mc^2$ states that the energy (E) which can be released from matter is equal to the mass (m) multiplied by the square of the speed of light (c).

Other mathematical formulae you may have encountered are:

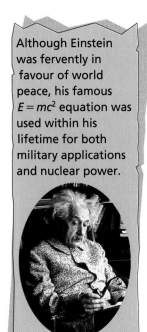

Although Einstein was fervently in favour of world peace, his famous $E = mc^2$ equation was used within his lifetime for both military applications and nuclear power.

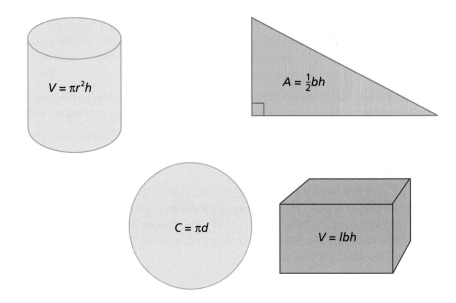

$V = \pi r^2 h$

$A = \frac{1}{2}bh$

$C = \pi d$

$V = lbh$

Substitution

$p + 2q - 5r$ is an **algebraic expression**.
$p + 2q = 5r$ is an **algebraic equation**.
$p = 5r - 2q$ is an **algebraic formula**.

It is important that you clearly understand the difference. Equations and formulae must have 'equals' signs.

Numbers can be **substituted** for letters in both expressions and formulae.

A formula is a general rule which enables the calculation of the value of one variable given the values of the other variables in the formula.

Example   Find the area of a triangle of base 8 cm and perpendicular height 6 cm.

$A = \frac{1}{2}bh$ is the general formula.
$A = \frac{1}{2} \times 8 \times 6$ is the equation for the above example.
$A = 24\,\text{cm}^2$

Substitution into expressions

Examples   Evaluate the expressions below if $a = 3$, $b = 4$ and $c = -5$.

$2a + 3b - c$
$= (2 \times 3) + (3 \times 4) - (-5)$
$= 6 + 12 + 5$
$= 23$

$a^2 - c^2$
$= 3^2 - (-5)^2$
$= 9 - 25$
$= -16$

$3a(2b - 3c)$
$= (3 \times 3) \times ((2 \times 4) - (3 \times -5))$
$= 9(8 - (-15))$
$= 9 \times 23$
$= 207$

Exercise 7.1

Evaluate the following expressions if $p = 2$, $q = 3$ and $r = 5$.

1 a $3p + 2q$ **b** $4p - 3q$ **c** $p - q - r$ **d** $3p - 2q + r$
2 a $-q(p + q)$ **b** $-2r(p - q)$ **c** $-3p(p - 3r)$ **d** $-4q(q - r)$
3 a $p^2 + q^2$ **b** $q^2 + r^2$ **c** $2p^2 - 3q^2$ **d** $3r^2 - 2q^2$

Evaluate the following expressions if $p = 4$, $q = -2$, $r = 3$ and $s = -5$.

4 a $2p + 4q$ **b** $5r - 3s$ **c** $3q - 4s$ **d** $6p - 8q + 4s$
   e $3r - 3p + 5q$ **f** $-p - q + r + s$
5 a $2p - 3q - 4r + s$ **b** $3s - 4p + r + q$ **c** $p^2 + q^2$ **d** $r^2 - s^2$
   e $p(q - r + s)$ **f** $r(2p - 3q)$
6 a $2s(3p - 2q)$ **b** $pq + rs$ **c** $2pr - 3rq$ **d** $q^3 - r^2$
   e $s^3 - p^3$ **f** $r^4 - q^5$

Substitution into formulae

The perimeter (P) of a rectangle is the distance around it.
For the rectangle on the right, the perimeter can be expressed as

$$l + b + l + b$$
$$\text{or } 2l + 2b$$
$$\text{or } 2(l + b)$$

> **Remember:**
> *To factorise an expression, look for a number or letter which divides into each term exactly.*

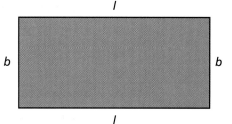

This can be written in the form $P = 2(l + b)$.
     The area (A) of a rectangle can be written in the form $A = lb$.

Exercise 7.2

1 Calculate the area and perimeter of rectangles of length *l* and breadth *b* as given below. Make sure your units are clearly written.

	length (*l*)	breadth (*b*)	perimeter	area
a	4 cm	7 cm		
b	8 cm	12 cm		
c	4.5 cm	2 cm		
d	8 cm	2.25 cm		
e	0.8 cm	40 cm		
f	1.2 cm	0.5 cm		
g	45 cm	1 m		
h	5.8 m	50 cm		

2 Find the area of each of the following rectangles:
   a the floor of a room measuring 6.5 m long by 5 m wide,
   b a sports hall floor which is 85 m long and 55 m wide,
   c the lid of a CD case which is a square of side length 135 mm,
   d a chessboard of side length 60 cm.

> **Remember:**
> *The perpendicular height is the height of the triangle measured at right angles to the base.*

The formula for the area of a triangle is given by $A = \frac{1}{2}bh$, where *A* represents the area of the triangle, *b* is the length of the base and *h* is the perpendicular height.

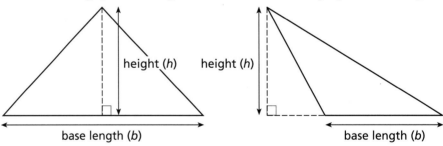

Exercise 7.3

1 Calculate the area of each of the following triangles.

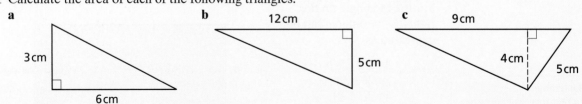

2 In physics the formula $V = IR$ is used in electricity, where

> V is the voltage in a circuit in volts
> I is the current in amps
> R is the resistance in ohms

Without using a calculator, calculate the voltage V when the current and resistance are as shown.

	current (*I*)	resistance (*R*)
a	7 amps	60 ohms
b	8 amps	400 ohms
c	0.3 amps	2000 ohms
d	80 milliamps	5000 ohms

> The terms Celsius and Centigrade are interchangeable.

The formula connecting a temperature in degrees Celsius (C) with a temperature in degrees Fahrenheit (F) is:

$$C = \tfrac{5}{9}(F - 32)$$

Example What is the Celsius equivalent of 77 °F?

$$C = \tfrac{5}{9}(F - 32)$$
$$C = \tfrac{5}{9}(77 - 32)$$
$$C = \tfrac{5}{9} \times 45$$
$$C = 25$$

Therefore 77 °F is equivalent to 25 °C.

Exercise 7.4

1 What is the Celsius equivalent of the following temperatures given in degrees Fahrenheit?
  a 59 °F   **b** 104 °F
  c 5 °F   **d** 212 °F
  e 32 °F

2 The highest temperature created by humans is 950 million °F. It was achieved at the Tokamak Fusion Test Reactor in the USA on 27 May 1994. Using the formula for converting degrees Fahrenheit to degrees Celsius, calculate the value of the highest temperature in degrees Celsius.

3 The lowest temperature possible is −459.67 °F. Calculate this temperature in degrees Celsius.

The circumference (C) of a circle of radius r is given by the formula:

$$C = 2\pi r$$

Example Calculate the circumference of a circle of radius 8 cm.

$$C = 2\pi r$$
$$C = 2 \times \pi \times 8$$
$$C = 50.3 \text{ cm (to 1 d.p.)}$$

Exercise 7.5

1 Calculate the circumferences of circles of radius given below. Give your answers correct to two decimal places.

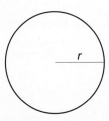

r

a $r = 6\,\text{cm}$ **b** $r = 18.5\,\text{cm}$ **c** $r = 0.5\,\text{cm}$
d $r = 1.1\,\text{cm}$ **e** $r = \pi\,\text{cm}$

2 The 'London Eye' wheel has a radius of 67.5 m. Calculate the total distance travelled by a person during one complete revolution of the wheel.

3 a The radius of a racing bike wheel is 55 cm. Calculate the circumference of the wheel.
  b During a 1 km race, how many *complete* rotations would the wheel make?

4 The formula for the area of a circle is given as $A = \pi r^2$, where A is the area of the circle and r its radius. Calculate the areas of circles of radius given below. Give your answers correct to two decimal places.

a $r = 4\,\text{cm}$ **b** $r = 9.5\,\text{cm}$ **c** $r = 18.3\,\text{cm}$
d $r = 0.8\,\text{cm}$ **e** $r = \pi\,\text{cm}$

Changing the subject of a formula

> **Remember:**
> *This is called **changing the subject** of the formula.*

In the formula $a = 2b + c$, 'a' is the **subject**. To make either b or c the subject, the formula has to be rearranged.

Examples   Rearrange the following formulae to make the letter in brackets the subject:
    a $l = 2m + n$ (n) **b** $2r + p = q$ (p) **c** $st = uv$ (u)

    a $l = 2m + n$ subtract $2m$ from both sides
        $l - 2m = n$
    b $2r + p = q$ subtract $2r$ from both sides
        $p = q - 2r$
    c $st = uv$ divide both sides by v

        $\dfrac{st}{v} = u$

The formula for the area of a circle is $A = \pi r^2$, where A is the area, and r represents the radius of the circle. If the area of the circle is 100 cm², calculate (to 2 d.p.) the circle's radius.

The formula $A = \pi r^2$ will need to be rearranged to make r the subject.

$$A = \pi r^2 \qquad \text{divide both sides by } \pi$$

$$r^2 = \frac{A}{\pi} \qquad \text{square root both sides}$$

Therefore $r = \sqrt{\dfrac{A}{\pi}}$

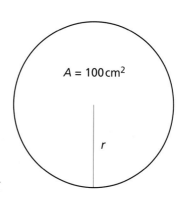

$A = 100\,\text{cm}^2$

r

If $A = 100\,\text{cm}^2$, then $r = \sqrt{\dfrac{100}{\pi}}$

$r = 5.64\,\text{cm}$

Exercise 7.6

In questions 1–4, make the letter in brackets the subject of the formula.

1 a $p + q = r$ (q)   **b** $q + 2r = s$ (q)   **c** $2q + r = 4p$ (r)   **d** $3s + q = 2p$ (q)

2 a $pq = r$ (q)   **b** $pr = qs$ (r)   **c** $pq = r + 3$ (p)   **d** $pr = q - 4$ (r)

3 a $m + n = r$ (n)   **b** $m + n = p$ (m)   **c** $2m + n = 3p$ (n)   **d** $3x = 2p + q$ (q)

  e $xy = uv$ (x)   **f** $pq = rs$ (s)

4 a $6q = 2p - 5$ (q)   **b** $6q = 2p - 5$ (p)   **c** $3x - 7y = 4z$ (z)   **d** $3x - 7y = 4z$ (x)

  e $3x - 7y = 4z$ (y)   **f** $2pr - q = 8$ (p)

5 The distance travelled by a boat moving at constant speed is given by the formula $d = st$, where d is the distance travelled (in metres), s is the speed travelled (in m/s) and t is the time spent travelling (in seconds).

  a Calculate the distance travelled if the boat was moving at 8 m/s for 60 seconds.

  b How long would it take the boat to travel 4000 m at 8 m/s?

  c What speed would the boat be travelling at if it travelled 6 km in 20 minutes?

6 The formula for the volume V of a cuboid of length l, breadth b and height h is given as $V = lbh$.

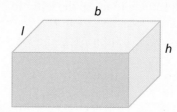

  a Calculate the volume of the cuboid if $l = 5$ cm, $b = 8$ cm and $h = 2$ cm.

  b Calculate the height of the cuboid if $V = 216$ cm³, $l = 8$ cm and $b = 9$ cm.

SUMMARY

By the time you have completed this chapter you should know:

- the difference between an **algebraic expression** and an **algebraic equation** or **formula**, for example $2a + 3b + 4c$ is an expression, $V = IR$ is a formula, $V = 7 \times 50$ (where $I = 7$ amps and $R = 50$ ohms) is an equation
- how to **substitute** numbers for letters in expressions, for example if $p = 5$, $q = -2$ and $r = 3$, the value of $2p + 3q - 4r$ by substitution is

$$(2 \times 5) + (3 \times -2) - (4 \times 3) = 10 - 6 - 12 = -8$$

- how to perform calculations that involve substituting numbers for letters in equations or formulae
- how to manipulate equations or formulae to **change the subject** of the formula, for example, if

$$p = \frac{r - sq}{t}, \text{ then } s = \frac{r - pt}{q}.$$

Exercise 7A

1 If $a = 3$, $b = 4$ and $c = 5$, evaluate the following expressions.
a $a + b + c$ **b** $4a - 3b$
c $a^2 + b^2 + c^2$ **d** $(a + c)(a - c)$

2 If $a = 2$, $b = 3$ and $c = 5$, evaluate the following expressions.
a $a - b - c$ **b** $2b - c$
c $a^2 - b^2 + c^2$ **d** $(a + c)^2$

3 Rearrange each of the following formulae to make the letter in brackets the subject.
a $p + q = r$ (p) **b** $r = q - s$ (q)
c $pr = qs$ (r) **d** $3s - t = 2r$ (s)

4 Rearrange each of the following formulae to make the letter in brackets the subject.
a $f - g = h$ (f) **b** $2h = g - 3k$ (g)
c $fk = gh$ (h) **d** $l = 5k - 3h$ (k)

Exercise 7B

1 If $x = 2$, $y = -3$ and $z = 4$, evaluate the following expressions.
a $2x + 3y - 4z$ **b** $10x + 2y^2 - 3z$
c $z^2 - x^2$ **d** $(z + x)(z - x)$

2 If $a = 4$, $b = 3$ and $c = -2$, evaluate the following expressions.
a $3a - 2b + 3c$ **b** $5a - 3b^2$
c $a^2 + b^2 + c^2$ **d** $(a + b)(a - b)$

3 Rearrange each of the following formulae to make the letter in brackets the subject.
a $x = 3p + q$ (q) **b** $3m - 5n = 8r$ (n)

4 Rearrange each of the following formulae to make the letter in brackets the subject.
a $p = 4m + n$ (n) **b** $4x - 3y = 5z$ (y)

Exercise 7C

Below is a sequence of matchstick patterns:

pattern 1:   number of matchsticks is 3

pattern 2:   number of matchsticks is 5

pattern 3:   number of matchsticks is 7

1 Draw the next two matchstick patterns and count the number of matchsticks used in each.
2 Copy and complete the table below, linking the number of triangles and the number of matchsticks needed.

number of triangles (n)	1	2	3	4	5
number of matchsticks (m)					

3 Write down an equation, with m as the subject, linking the number of triangles (n) to the number of matchsticks (m).
4 Use your equation to work out the total number of matchsticks needed for a pattern of 50 triangles.
5 Rearrange your equation to make n the subject.
6 How many triangles would there be in a pattern consisting of 161 matches?
7 Investigate matchstick patterns made of square shapes, for example

pattern 4    13 matchsticks

8 Investigate matchstick patterns made of pentagon shapes, for example

pattern 3 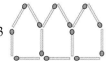   13 matchsticks

9 Try to write a general equation linking the number of matches (m) to the number of shapes (n) each with number of sides (s).
10 Using your equation from question 9, calculate the number of matches needed for a pattern made up of 15 *octagonal* shapes.

Exercise 7D

You will need:
- A4 paper
- compasses
- computer with a spreadsheet package installed

1 On a piece of A4 paper, draw eight different-sized rectangles. Measure each rectangle's length and width and label them clearly on your diagrams.

2 On a piece of A4 paper, draw eight different-sized circles. Mark and measure the radius of each circle.

3 On a spreadsheet (see example below) enter the dimensions of your rectangles and circles in two separate tables. Write formulae so that the spreadsheet automatically calculates the area of each of your rectangles and circles.

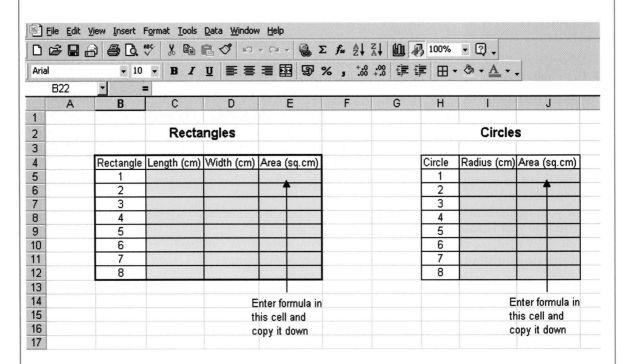

Exercise 7E

Isaac Newton was a great mathematician who had done important work by the age of 23. This included work on rainbows, integration and the motion of the Moon. Find out more about these three subjects, with formulae where possible.

8 Sequences

The pyramid is a shape which has fascinated people for thousands of years. In Egypt, a pyramid was a burial place for Pharaohs and, in South America, pyramids were temples. Pyramids have also been found in ancient cities in Laos and Cambodia.

The triangular faces of the pyramids were formed from stone blocks. These could have been arranged in a variety of ways – two possible arrangements are shown below:

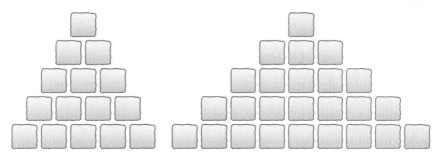

The successive total number of blocks as each level is included would be:

1	1
3	4
6	9
10	16

Each of these is called a **sequence** of numbers.

A **number sequence** is an ordered set of numbers. Each number in the sequence is called a **term**. The terms of a sequence form a pattern, for example:

$$1 \quad 2 \quad 4 \quad 8 \quad 16$$

This is a sequence in which we are doubling each of the terms to get the next term.

$$1 \quad 4 \quad 9 \quad 16 \quad 25$$

This is a sequence of successive square numbers.

Remember:
A sequence is a list of numbers which follow a pattern or rule.

Exercise 8.1

You will need:
• squared paper

1

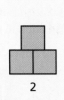 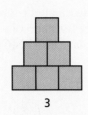

1   2   3

The above sequence of block patterns follows a particular rule.
a Draw the first six patterns in the above sequence.
b Copy and complete the table below linking the pattern number to the number of squares in each pattern.

pattern number	1	2	3	4	5	6
number of squares						

c How many squares would be needed to draw the tenth pattern?
d Without drawing the 50th pattern, explain how you would work out the number of squares needed to draw it.

2

  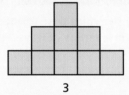

1   2   3

The above sequence of block patterns follows a particular rule.
a Draw the first five patterns in the above sequence.
b Copy and complete the table below linking the pattern number to the number of squares in each pattern.

pattern number	1	2	3	4	5
number of squares					

c Describe the sequence of numbers represented by the number of squares.
d How many squares would be needed to draw the 100th pattern? Explain clearly how you arrived at your answer.

3 King Ozymandias decided to have a monument in the form of a pyramid (similar to the one on the right) built as a lasting monument. It was to be made from blocks of stone shaped into cubes.

He was not sure how high to build his pyramid, so he asked his architects some questions.

a How many cubes are needed to build the pyramid above?

b How many cubes would be needed to build a pyramid 10 cubes high?

c Explain your method for solving part **b** above.

d Use the same method to complete the table below for pyramids of heights 1–10 cubes.

height	1	2	3	4	5	6	7	8	9	10
total number of blocks										

e Using the same method, how many blocks would be required to build a pyramid 100 blocks high?

f If Ozymandias' architects had 1000 blocks of stone, what would be the highest pyramid of this shape that they could build?

4

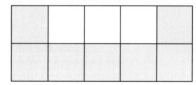

The above sequence of block patterns follows a particular rule.

a Draw the first five patterns in the above sequence.

b Copy and complete the table below linking the number of white squares to the number of coloured squares in each pattern.

number of white squares	1	2	3	4	5
number of coloured squares					

c Describe the relationship between the numbers of white and coloured squares.

d If a pattern is made of 1000 coloured squares, calculate the number of white squares needed.

Term-to-term rules

A rule which describes how to get from one term to the next is called a **term-to-term rule**.

Examples

Below is a sequence of numbers.

$$5 \quad 9 \quad 13 \quad 17 \quad 21 \quad 25$$

a What is the term-to-term rule for the sequence?
b Calculate the 10th term of the sequence.

a The term-to-term rule is 'add 4' (+4).
b Continuing the pattern gives:

$$5 \quad 9 \quad 13 \quad 17 \quad 21 \quad 25 \quad 29 \quad 33 \quad 37 \quad 41$$

Therefore the 10th term is 41.

Below is a sequence of numbers.

$$1 \quad 2 \quad 4 \quad 8 \quad 16$$

a What is the term-to-term rule for the sequence?
b Calculate the 10th term of the sequence.

a The term-to-term rule is ×2.
b Continuing the pattern gives:

$$1 \quad 2 \quad 4 \quad 8 \quad 16 \quad 32 \quad 64 \quad 128 \quad 256 \quad 512$$

Therefore the 10th term is 512.

Remember to look very carefully. Consider the sequence

$$1 \quad 2 \quad 4 \quad \ldots \quad \ldots$$

The next term could be 8 (if doubling each time) or it could be 7 (if successive differences are $1, 2, 3, \ldots$).

Exercise 8.2

For each of the sequences given below, give in words a rule to describe the sequence, and calculate the 10th term.

a	3	6	9	12	15
b	8	18	28	38	48
c	11	33	55	77	99
d	0.7	0.5	0.3	0.1	
e	$\frac{1}{2}$	$\frac{1}{3}$	$\frac{1}{4}$	$\frac{1}{5}$	
f	$\frac{1}{2}$	$\frac{2}{3}$	$\frac{3}{4}$	$\frac{4}{5}$	
g	1	4	9	16	25
h	4	7	12	19	28
i	1	8	27	64	
j	2	9	28	65	

Method of differences

Sometimes the pattern in a sequence of numbers is not obvious. But, by looking at the differences between successive terms, it is often possible to find a pattern. Constructing a **table of differences** can help you find the pattern.

Examples Calculate the eighth term in the sequence below.

          8   12   20   32   48

The pattern in the sequence above is not immediately obvious, so a table can be constructed showing the differences between successive terms:

                      8   12   20   32   48
first differences   4   8   12   16

The pattern in the differences row is +4 and this can be continued to complete the sequence to the eighth term:

                      8   12   20   32   48   68   92   120
first differences   4   8   12   16   20   24   28

Calculate the eighth term in the sequence below.

          3   6   13   28   55

Construct a table of differences.

                      3   6   13   28   55
first difference   3   7   15   27

If the row of first differences is not sufficient to spot the pattern, then a row of second differences can be constructed.

                      3   6   13   28   55
first difference   3   7   15   27
second difference   4   8   12

The pattern in the second differences row can be seen to be +4. This can now be used to complete the sequence.

                      3   6   13   28   55   98   161   248
first difference   3   7   15   27   43   63   87
second difference   4   8   12   16   20   24

Exercise 8.3

For each of the sequences given below, calculate the next two terms.

a	8	11	17	26	38	
b	5	7	11	19	35	
c	9	3	3	9	21	
d	−2	5	21	51	100	
e	11	9	10	17	36	79
f	4	7	11	19	36	69
g	−3	3	8	13	17	21	24

The *n*th term

So far the method used for generating a sequence relies on knowing the previous term in order to work out the next one. This method works, but can be a little cumbersome if the 100th term is needed and only the first five terms are given! A more efficient rule is one which relates a term to its position in the sequence.

Examples   For the sequence shown below, give an expression for the *n*th term.

position	1	2	3	4	5	*n*
term	3	6	9	12	15	?

By looking at the sequence it can be seen that the term is always $3 \times$ its position. Therefore the *n*th term can be given by the expression $3n$.

For the sequence shown below, give an expression for the *n*th term.

position	1	2	3	4	5	*n*
term	2	5	8	11	14	?

You may have spotted the similarities between these sequences. The terms of the above sequence are the same as the terms in the first example above, but with 1 subtracted each time.
The expression for the *n*th term is therefore $3n - 1$.

Exercise 8.4

For each of the following sequences write down the next two terms, and give an expression for the *n*th term.

a 5	8	11	14	17		
c 4	9	14	19	24		
e 1	8	15	22	29		
g 1	10	19	28	37		
i 9	20	31	42	53		
k 0.25	1.25	2.25	3.25	4.25		

b 5	9	13	17	21		
d 8	10	12	14	16		
f 0	4	8	12	16	20
h 15	25	35	45	55		
j 1.5	3.5	5.5	7.5	9.5	11.5
l 0	1	2	3	4	5

Exercise 8.5

For each of the following sequences write down the next two terms, and give an expression for the *n*th term.

a 2	5	10	17	26	37	
c 0	3	8	15	24	35	
e 2	9	28	65	126		
g −2	5	24	61	122		

b 8	11	16	23	32		
d 1	8	27	64	125		
f 11	18	37	74	135		
h 2	6	12	20	30	42

Generating sequences

If we know the general rule for the nth term of a sequence, then we can generate the sequence using the rule, because it enables us to calculate individual terms.

Example   The nth term of a sequence is given by the rule $2n + 1$.
Calculate the first five terms of the sequence.

When $n = 1$:
$$2n + 1 = (2 \times 1) + 1 = 3$$

When $n = 2$:
$$2n + 1 = (2 \times 2) + 1 = 5$$

When $n = 3$:
$$2n + 1 = (2 \times 3) + 1 = 7$$

When $n = 4$:
$$2n + 1 = (2 \times 4) + 1 = 9$$

When $n = 5$:
$$2n + 1 = (2 \times 5) + 1 = 11$$

Entering the results into a table gives:

position	1	2	3	4	5
term	3	5	7	9	11

Exercise 8.6

For each of the rules for the nth term given below, give the value of the first five terms in the sequence, and the twentieth term. Present your solutions in a table.

1 a $3n + 1$ **b** $6n - 3$ **c** $5n - 4$ **d** $3 - 2n$ **e** $5 - 4n$

2 a $n^2 + 1$ **b** $n^2 - 3$ **c** $5 - n^2$ **d** $n^3 + 10$ **e** $8 - n^3$

3 a $\dfrac{n}{2} + 1$ **b** $\dfrac{n}{3} - 1$ **c** $\dfrac{n + 4}{2}$ **d** $\dfrac{2n - 5}{5}$

SUMMARY

By the time you have completed this chapter you should know:

■ what is meant by a **number sequence** – an ordered set of numbers which may form a pattern, for example 3, 6, 9, 12, 15, . . .;   1, 4, 9, 16, 25, . . .;   5, 11, 17, 23, 29, . . .
Each number in a sequence is called a **term**

■ that a **term-to-term rule** describes how to get from one term to the next

■ how to draw up and use a **table of differences** to help find the next term in a sequence; for example, to calculate the sixth term in the sequence below:

$$
\begin{array}{ccccccc}
 & 7 & 11 & 19 & 31 & 47 & ? \\
\text{first difference} & & 4 & 8 & 12 & 16 & 20 \\
\text{second difference} & & & 4 & 4 & 4 & 4
\end{array}
$$

■ how to construct a rule for the ***n*th term** in a sequence, for example, to find the *n*th term for the sequence of numbers 1, 4, 7, 10, 13, 16, by looking at the sequence, you will be able to work out that the *n*th term is $3n - 2$; but you must *always* check that the formula works.

Exercise 8A

You will need:
• squared paper

1
 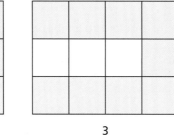

      1   2   3

  a Draw the next two patterns in the above sequence.
  b Copy and complete the table below linking the number of white squares to the number of coloured squares.

number of white squares	1	2	3	4	5
number of coloured squares					

  c Write the rule for the *n*th term of the sequence.
  d Use your rule to predict the number of coloured squares in a pattern with 50 white squares.
2 For each of the sequences given below, calculate the next two terms, and explain the pattern in words.
  a   9   18   27   36
  b 54   48   42   36
  c 18   9   4.5
  d 12   6   0   −6
  e 216   125   64
  f   1   3   9   27

3 For each of the sequences shown below, give an expression for the *n*th term.

   a 6   10   14   18   22

   b 13   19   25   31

   c 3   9   15   21   27

   d 4   7   12   19   28

   e 0   10   20   30   40

   f 0   7   26   63   124

Exercise 8B

You will need:
• squared paper

1

       1   2   3

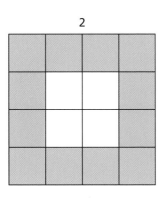
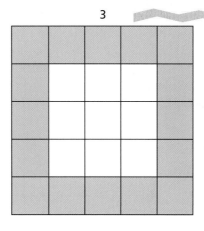

   a Draw the next two patterns in the above sequence.

   b Copy and complete the table below linking the pattern number to the number of white squares and the number of coloured squares.

pattern number	1	2	3	4	5
number of white squares					
number of coloured squares					

   c Write the rule for the *n*th term of the number of white squares.

   d Write the rule for the *n*th term of the number of coloured squares.

   e Use your rules to predict the number of white and coloured squares needed for the 20th pattern.

2 For each of the sequences given below, calculate the next two terms, and explain the pattern in words.

   a 6   12   18   24

   b 24   21   18   15

   c 10   5   0

   d 16   25   36   49

   e 1   10   100

   f 1 $\frac{1}{2}$ $\frac{1}{4}$ $\frac{1}{8}$

3 For each of the sequences shown below, give an expression for the *n*th term.

   a 3   5   7   9   11

   b 7   13   19   25   31

   c 8   18   28   38

   d 1   9   17   25

   e −4   4   12   20

   f 2   5   10   17   26

Exercise 8C

You will need:
- pair of compasses
- ruler

The diagram below is described as a twelve-point geometric rose. It has been drawn by connecting all 12 points on the circumference of the circle to each other with straight lines. Therefore, in the complete diagram, each point is connected to every other point.

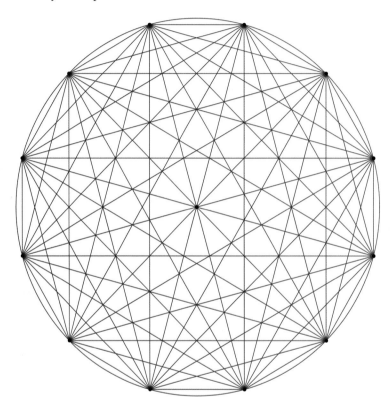

By drawing your own geometric roses, answer the following questions.

1 How many lines are there in a four-point geometric rose?
2 How many lines are there in a five-point geometric rose?
3 How many lines are there in a six-point geometric rose?
4 Investigate the total number of straight lines in geometric roses.

Exercise 8D

The expressions for the *n*th term of six sequences are given below:

$$2n+1 \qquad n^2-3 \qquad \tfrac{1}{2}n-4 \qquad 3n+2 \qquad 2n^2+1 \qquad \tfrac{1}{2}n^2+3$$

Use a spreadsheet with formulae to generate the first 10 terms of each sequence. You could set your spreadsheet out in a similar way to the one shown below:

You will need:
- computer with a spreadsheet package installed

File Edit View Insert Format Tools Data Window Help

Arial   10   **B** *I* U

D21   =

	A	B	C	D	E	F	G	H	I	J
1										
2		**2n + 1**			**n^2 - 3**			**0.5n - 4**		
3		Position	Term		Position	Term		Position	Term	
4		1			1			1		
5		2			2			2		
6		3			3			3		
7		4			4			4		
8		5			5			5		
9		6			6			6		
10		7			7			7		
11		8			8			8		
12		9			9			9		
13		10			10			10		
14										

Using the graphing facility of the spreadsheet, draw line graphs for each sequence. What can you deduce about the shape of the graph and the formula for the *n*th term?

Exercise 8E

Leonardo of Pisa was an Italian mathematician born in the twelfth century. By what nickname do we know him, and for which sequence of numbers is he famous?

9 Linear graphs

René Descartes (1596–1650) was a French philosopher and mathematician. He is considered one of the most original thinkers of all time. His greatest work is *The Meditations* (published in 1641) which asked 'How and what do I know?'. His work in mathematics formed a link between algebra and geometry. He believed that mathematics was the supreme science in that the whole phenomenal world could be interpreted in terms of mathematical laws.

Equation of a straight line

A line is made up of an infinite number of points. The coordinates of every point on a straight line all have a common relationship. In other words, the x- and y-values follow a pattern.

The line below is plotted on a pair of axes.

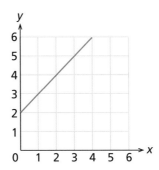

By looking at the coordinates of some of the points on the line, we can see a pattern:

x	0	1	2	3	4
y	2	3	4	5	6

In words, the pattern linking the x- and y-coordinates can be described as: the y-coordinates are 2 more than the x-coordinates. Written algebraically, this is $y = x + 2$. This is known as the **equation of the straight line**.

Examples   By looking at the coordinates of some of the points on the line below, find the equation of the straight line.

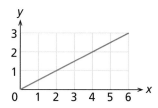

x	0	2	4	6
y	0	1	2	3

By looking at the table, it can be seen that the rule linking each pair of coordinates is that the y-coordinate is half the value of the x-coordinate. Therefore $y = \frac{1}{2}x$.

By looking at the coordinates of some of the points on the line below, find the equation of the straight line.

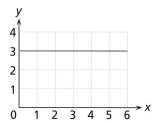

x	0	1	2	3
y	3	3	3	3

By looking at the table, it can be seen that the only rule that all the points have in common is that the y-values are always equal to 3. Therefore the equation of the straight line is $y = 3$.

Exercise 9.1

In each of the following, identify the coordinates of some of the points on the line and use these to find the equation of the straight line.

1

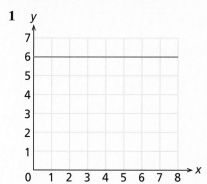

2

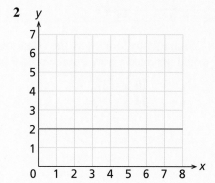

3
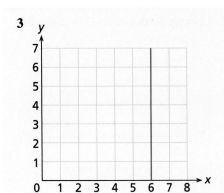

4
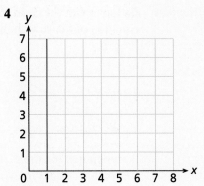

5
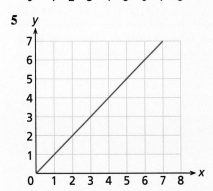

6
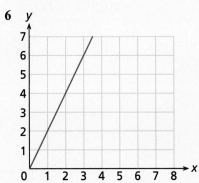

7
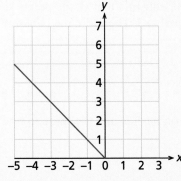

8
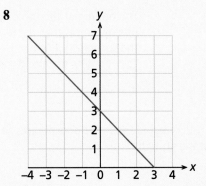

For questions 9–15, simply by looking at the equation, work out whether, if points were plotted for this equation, they would produce a horizontal, vertical or diagonal line.

9 $y = 6$ **10** $x = 3$ **11** $y = x + 3$ **12** $y = x - 2$

13 $x = 6$ **14** $y = -3$ **15** $y + x = 5$

Drawing straight-line graphs

Remember:
It is good practice to plot a third point as a check.

To draw a straight-line graph only two points need to be known. Once these have been plotted, the line can be drawn between them and extended if necessary at both ends.

Example Plot the line $y = x + 3$.

To identify two points, simply choose two values for x. Substitute these into the equation and calculate their corresponding y-values.
    When $x = 0$, $y = 3$.
    When $x = 4$, $y = 7$.
    Therefore two of the points on the line are $(0, 3)$ and $(4, 7)$. The straight line $y = x + 3$ is plotted below.

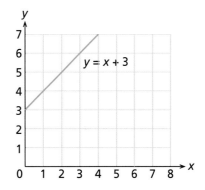

Check a third point:
    when $x = 3$, $y = 6$

As the point $(3, 6)$ lies on the line, the graph is correct.

Exercise 9.2

You will need:
• squared paper

Plot the following straight lines.

1 $y = 2x + 3$ **2** $y = x - 4$ **3** $y = 3x - 2$
4 $y = -2x$ **5** $y = -x - 1$ **6** $y - 4 = 3x$

Straight-line graphs in real life

Many situations in real life generate straight-line graphs. Currency conversions are a good example of this.

£1 is approximately equivalent to 9 French francs. The table below gives the conversions for the range £1–£5:

£ sterling (x)	1	2	3	4	5
French francs (y)	9	18	27	36	45

If x represents the number of pounds sterling and y the number of French francs, then the relationship between x and y can be written as an equation in the form $y = 9x$. When plotted, the relationship between x and y can clearly be seen as linear.

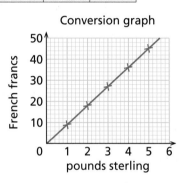

Examples Using the graph on the right, estimate the number of French francs equivalent to £3.50.

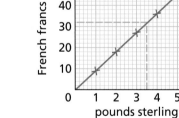

£3.50 is therefore approximately equivalent to 32 francs.

Using the graph, estimate the number of pounds sterling equivalent to 60 francs.

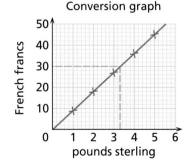

The graph does not go up to 60 francs. However, it can still be used. 30 francs can be seen to be approximately equivalent to £3.30, therefore 60 francs will be approximately equivalent to £6.60.

Because we are dealing with money, we read values off as £3.30 not £3.3.

Exercise 9.3

You will need:
• graph paper

1 £1 is equivalent to $1.60.
   a Copy and complete the conversion table below.

£ (x)	0	1	2	3	4	5
$ (y)		1.60				

   b If x represents the number of pounds sterling and y the number of dollars, express the relationship between x and y as an equation.
   c Plot a graph to show the relationship between x and y.
   d Use your graph to estimate the number of dollars equivalent to £4.25.
   e Use your graph to estimate the number of pounds sterling equivalent to $24.

2 A DIY store sells fencing panels at £25 each.
   a Copy and complete the table below.

number of panels (n)	1	2	3	4	5
cost in £ (C)		50			

   b If n represents the number of panels and C the cost in pounds, express the relationship between n and C as an equation.
   c The DIY store also offers a delivery service. It charges a flat rate of £10, irrespective of the number of panels bought. Copy and complete the table below, showing the cost of buying panels if they are delivered.

number of panels (n)	1	2	3	4	5
cost in £ (C)		60			

   d Express the relationship between n and C as an equation.
   e Using the same axes, plot the results obtained in parts a and c above. Label your two lines clearly.
   f Use your graph to work out how many *complete* panels could be bought for £100 if:
      i) they were not delivered,
      ii) they were delivered.

3 A stone is thrown downwards from the edge of a cliff. Its speed v (m/s) at a time t (seconds) is given by the equation:

$v = 10t + 8$

   a Using the above equation, copy and complete the table below.

t (seconds)	0	1	2	3	4	5
v (m/s)			28			

   b Use your table of results to plot a graph of t against v.
   c Use your graph to estimate the speed of the stone after 8 seconds.
   d Use your graph to estimate the time taken to reach a speed of 50 m/s.

4 'Tangerine', a mobile phone company, charges a line rental fee of £15 a month and then a flat rate of 5p per unit used.

a Copy and complete the table of monthly charges below.

number of units (n)	0	100	200	300	400	500
charge in £ (C)						

b If n represents the number of units used and C the resulting charge in £, express the relationship between n and C as an equation.

c Plot a graph to show the relationship between n and C.

d Use your graph to estimate the cost of using 220 units.

e Use your graph to estimate how many units were used if the total charge was £30.25.

Simultaneous equations

You saw earlier in the chapter how to draw straight-line graphs from a linear equation by identifying two points on the line (with a third as a check). If two linear equations plotted on the same pair of axes intersect, then the point of intersection is the solution of both equations simultaneously. Hence the term **simultaneous equations**.

Example **a** Plot the graphs given by these two simultaneous equations:

$$y = x + 3 \qquad y = 5 - x$$

b From the point of intersection give values of x and y which solve the simultaneous equations above.

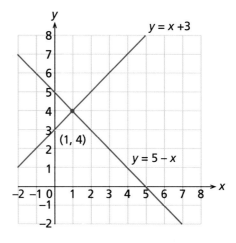

The graphs intersect at the point (1, 4) so the solution to the simultaneous equations is

$$x = 1 \text{ and } y = 4$$

Simultaneous equations can also be solved algebraically.

Example A mobile phone company has two offers.

Offer A: a line rental fee of £10 and an additional 5p per unit used
Offer B: no line rental, but a charge of 8p per unit used

a Write an equation linking the total charge C and the number of units U, for each offer.

b Copy and complete the table of charges below for each of the offers.

numbers of units (U)	0	100	200	300	400	500
offer A charge £ (C)	10					
offer B charge £ (C)	0					

c Using the same axes, plot the graphs for both offers.

d Which offer is cheaper if a customer uses a small number of units?

e Use the graph to estimate the number of units that need to be used before the other offer becomes cheaper.

a Offer A: $C = 10 + 0.05U$
 Offer B: $C = 0.08U$

b

number of units (U)	0	100	200	300	400	500
offer A charge £ (C)	10	15	20	25	30	35
offer B charge £ (C)	0	8	16	24	32	40

c

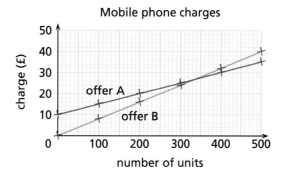

Mobile phone charges

d Offer B is cheaper.

e This is found at the **point of intersection** of the two lines. Therefore the number of units that need to be used before offer A becomes cheaper is approximately 330 units.

Remember:
When solving simultaneous equations by graphs, the point of intersection provides the solution.

Exercise 9.4

You will need:
• graph paper

1 A family wishes to hire a car while on holiday. There are two car hire companies to choose from: 'Autohire' and 'Wheels 4U'. Their charges for similar cars are shown below:

   a Copy and complete the table of charges below, for each of the car hire companies.

number of days (*n*)	1	2	3	4	5	6	7
Autohire charge £ (C)	25						
Wheels 4U charge £ (C)	55						

   b If *n* represents the number of days a car is hired for and *C* is the resulting charge in pounds sterling, express the relationship between *n* and *C* as an equation, for each of the car hire companies.

   c Using the same axes, plot both graphs. Make sure that the lines are clearly labelled.

   d Use your graph to show which of the car hire companies the family should choose if they wish to hire a car for three days.

   e Which car hire company should the family choose if they wish to hire a car for four days? Use the graph to explain your answer.

2 Maria is playing a 'guess the number' game with Gabriella. Maria is thinking of two numbers *x* and *y*. Maria's two *x* and *y* numbers satisfy both the following relationships:

$$y = x + 6 \quad \text{and} \quad y = 3x - 2$$

   a Copy and complete the tables below for the two relationships:

$y = x + 6$

x	1	3	5	7	9
y					

$y = 3x - 2$

x	1	3	5	7	9
y					

   b On the same axes, plot and label the graphs of $y = x + 6$ and $y = 3x - 2$.

   c What does the point of intersection of both graphs represent?

   d What are the two numbers that Maria is thinking of?

Experimental results

A **line of best fit** is a line drawn on a graph which passes through or close to the maximum number of points plotted on a graph of experimental results, as below:

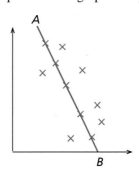

Line *AB* is the line which 'best fits' the results (if a straight line is expected by theory).

Exercise 9.5

You will need:
• graph paper

1 A car is travelling steadily along a straight road. Markers are placed at 1 km intervals and the time is recorded as the car passes each one. The table below shows the times registered for 5 km.

distance (m)	0	1000	2000	3000	4000	5000
time (seconds)	0	50	124	178	240	310

  a Using the values of the table to indicate how far you need to label your axes, plot the points shown.

  b Draw a line of best fit through the points.

  c Estimate, using your line of best fit, how far the car had travelled after 200 seconds.

  d If the car continued to travel at the same constant speed, as indicated by the line of best fit, estimate how far it would have travelled after 20 minutes.

> **Remember:**
> *A line of best fit is a straight line that goes as near as possible to all the points.*

2 When Cheryl's car was checked, it was found to have a faulty speedometer. The speedometer readings and the true values are recorded in the table below.

speedometer (km/h)	0	20	40	60	80	100
true speed (km/h)	0	20	45	74	100	120

  a Using the values in the table above to indicate how far you need to label your axes, plot the points shown.

  b Draw a line of best fit through the points.

  c Use the line of best fit to predict what Cheryl's true speed would be if her speedometer indicated a speed of 70 km/h.

SUMMARY

By the time you have completed this chapter you should know:

■ how to find the **equation of a straight line** by using the coordinates of some of the points on it; for example if a straight line has points PQRS on it, P (1, 4), Q (2, 5), R (3, 6), S (4, 7), the pattern linking them is that the y-coordinate is 3 more than the x-coordinate, so the equation is $y = x + 3$

■ how to draw straight-line graphs from an equation; for example, to plot the line represented by the equation $y = x + 4$, identify three points (the third is a check) by choosing values for x and substituting these in the equation to calculate the corresponding y-values

$x = 1, y = x + 4 = 1 + 4 = 5$ coordinates (1, 5)
$x = 2, y = x + 4 = 2 + 4 = 6$ coordinates (2, 6)
check: $x = 5, y = x + 4 = 5 + 4 = 9$ coordinates (5, 9)

then plot these coordinates on a graph and join them up for the line representing $y = x + 4$

■ that some real-life situations produce straight-line graphs, for example a currency conversion graph as below, where £1 = US$1.60

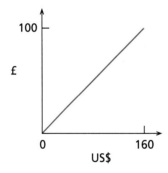

■ how to use straight-line graphs to give approximate solutions to real-life problems

■ the **point of intersection** when two lines are plotted on the same pair of axes provides the solution to two simultaneous equations, for example $y = 5 - x$ and $y = x - 1$

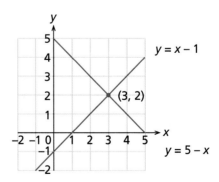

The graphs intersect at point (3, 2) so the solution to the simultaneous equations is $x = 3, y = 2$

■ that **lines of best fit** can be used to deduce results from experimental data.

Exercise 9A

You will need:
• squared paper

1

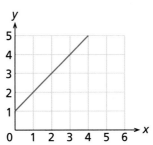

a Copy and complete the table below for the coordinates of some of the points on the line drawn above.

x	0	1	2	3	4
y					

b Using your table of results, work out the equation of the straight line.

2 a Copy the axes below.

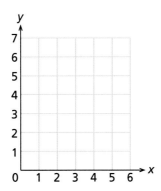

b On your axes, plot the line $y = 2x + 1$.

3 £1 is equivalent to 300 Portuguese escudos.

a Copy and complete the conversion table below.

£	1	2	3	4	5	6	7	8	9	10
esc.										

b Draw a conversion graph using the results from your table.
c Use your graph to estimate the number of Portuguese escudos equivalent to £4.50.
d Use your graph to estimate the number of pounds equivalent to 2475 escudos.

4 a Copy and complete the tables for the relationships below.

$y = 4 - 2x$ $\qquad\qquad\qquad\qquad y = \dfrac{7 - x}{3}$

x	0	1	2	3	4
y					

x	0	1	2	3	4
y					

b On the same axes, plot and label the graphs of $y = 4 - 2x$ and $y = \dfrac{7 - x}{3}$.

c Write down the coordinates of the point of intersection of the two lines.

Exercise 9B

You will need:
- squared paper
- graph paper

1

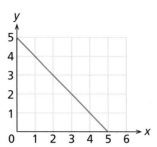

a Copy and complete the table below for the coordinates of some of the points on the line drawn above.

x	0	1	2	3	4
y					

b Using your table of results, deduce the equation of the straight line.

2 a Copy the axes below.

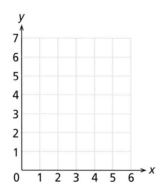

b On your axes plot the line $y = 6 - \frac{1}{2}x$.

3 £1 is equivalent to 170 Japanese yen.

a Copy and complete the conversion table below.

£	1	2	3	4	5	6	7	8	9	10
¥										

b Draw a conversion graph using the results from your table.

c Use your graph to estimate the number of yen equivalent to £3.75.

d Use your graph to estimate the number of pounds equivalent to 1400 yen.

4 a Copy and complete the tables for the relationships below.

$y = 3x - 1$ $y = x + 3$

x	0	1	2	3	4
y					

x	0	1	2	3	4
y					

b On the same axes, plot and label the graphs of $y = 3x - 1$ and $y = x + 3$.

c Write the coordinates of the point of intersection of the two lines.

Exercise 9C

The following investigation is a whole-class practical activity.

- Collect a number of cylindrical objects – for the best results you should try to collect a range of different-sized cylinders.
- For each cylinder, measure as accurately as possible the radius of one of the circular faces and the circumference.
- Enter your results into a spreadsheet as below:

| File | Edit | View | Insert | Format | Tools | Data | Window | Help |

| Arial | ▾ | 10 | ▾ | **B** | *I* | U | | | |

| D15 | ▾ | = |

	A	B	C	D	E
1					
2		**Object**	**Radius**	**Circumference**	
3					
4					
5					
6					
7					
8					
9					

- Using the graphing facility of the spreadsheet, plot a scatter graph of radius against circumference.
- What do you notice about the way the points are lying?
- Draw a line of best fit through your points.
- Use your line of best fit to predict the dimensions of other cylindrical objects.
- Extension: find the equation of the line drawn.

Exercise 9D

Use a graphical calculator to calculate the coordinates of the point of intersection of the following pairs of linear equations.

1 $y = 6 - x$ $y = x - 2$
2 $y = 11 - x$ $y = x - 1$
3 $y = 5 - x$ $y = x - 7$
4 $y = 12 - 2x$ $y = 2x - 8$
5 $y = 17 - 3x$ $y = 3x - 13$
6 $y = 29 - 5x$ $y = 5x - 11$

Exercise 9E

'Cogito, ergo sum' was a thought expressed by which mathematician/ philosopher? It is said that this man was 'the father of modern philosophy' – see if you can find out why.

Jostein Gaarder's book, *Sophie's World*, may help you.

10 Gradients

Remember:
On a travel graph, the gradient of the line tells you the speed of the object. The steeper the line, the faster the object is moving.

The **gradient** of a slope is an indication of how steep the slope is. In the real world, knowing how steep a slope is can be particularly useful when riding a bike or driving a car, or when skiing, or when simply walking up or down a hill.

Gradient of a straight line

The gradient of a straight line refers to its 'steepness' or 'slope'. A straight line has a **constant** gradient, i.e. it does not change. The gradient can be calculated by considering the coordinates of two points on the line.

Examples The coordinates of two points on a straight line are (1, 3) and (5, 7). Plot the two points on a coordinate grid and calculate the gradient of the line joining them.

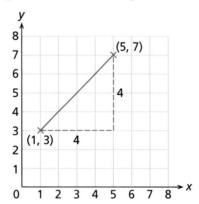

The gradient of the line joining two points is calculated using the following formula:

$$\text{gradient} = \frac{\text{vertical height between the two points}}{\text{horizontal distance between the two points}}$$

So the gradient is the difference between the y-coordinates, divided by the difference between the x-coordinates.

By considering the x- and y-coordinates of the two points, this can be rewritten as:

$$\text{gradient} = \frac{y_2 - y_1}{x_2 - x_1}$$

$$\text{gradient} = \frac{7 - 3}{5 - 1} = \frac{4}{4} = 1$$

Note. It does not matter which point we choose to be (x_1, y_1) or (x_2, y_2) as the gradient will still be the same. In the example above:

$$\text{gradient} = \frac{3 - 7}{1 - 5} = \frac{-4}{-4}$$

$$= 1$$

The coordinates of two points on a straight line are $(2, 6)$ and $(4, 2)$. Plot the two points on a coordinate grid and calculate the gradient of the line joining them.

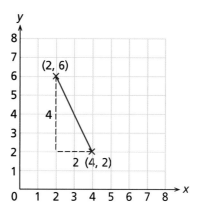

$$\text{gradient} = \frac{6 - 2}{2 - 4} = \frac{4}{-2}$$

$$= -2$$

To check whether or not the sign of the gradient is correct, the following guideline is useful:

a line sloping this way has a positive gradient   a line sloping this way has a negative gradient

Exercise 10.1

1 With the help of a coordinate grid if necessary, calculate the gradient of the line joining each of the following pairs of points.

  a $(5, 6), (1, 2)$ **b** $(6, 4), (3, 1)$ **c** $(1, 4), (5, 8)$

  d $(0, 0), (4, 8)$ **e** $(2, 1), (4, 7)$ **f** $(0, 7), (-3, 1)$

  g $(-3, -3), (-1, 5)$ **h** $(4, 2), (-4, -2)$ **i** $(-3, 5), (4, 5)$

  j $(2, 0), (2, 6)$ **k** $(-4, 3), (4, 5)$ **l** $(3, 6), (-3, -3)$

2 With the help of a coordinate grid if necessary, calculate the gradient of the line joining each of the following pairs of points.

  a $(1, 4), (4, 1)$ **b** $(3, 6), (7, 2)$ **c** $(2, 6), (6, -2)$

  d $(1, 2), (9, -2)$ **e** $(0, 3), (-3, 6)$ **f** $(-3, -5), (-5, -1)$

  g $(-2, 6), (2, 0)$ **h** $(2, -3), (8, 1)$ **i** $(6, 1), (-6, 4)$

  j $(-2, 2), (4, -4)$ **k** $(-5, -3), (6, -3)$ **l** $(3, 6), (5, -2)$

In the previous chapter we looked at how a straight line is represented by an equation. The equation describes the relationship between the *x*- and *y*-coordinates of all the points on the line. It can be deduced by looking at the coordinates of some of the points on the line.

Exercise 10.2

1 For each of the following, identify the coordinates of some of the points on the line and use these to find the equation of the straight line.

a

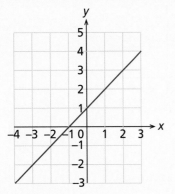

b

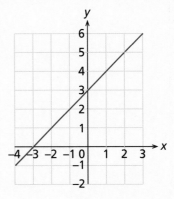

c

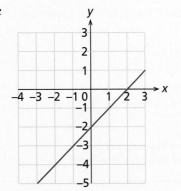

d

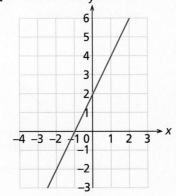

e

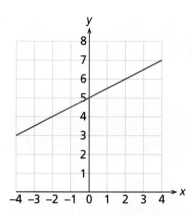

f

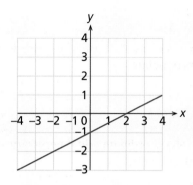

2 For each of the following, identify the coordinates of some of the points on the line and use these to find the equation of the straight line.

a

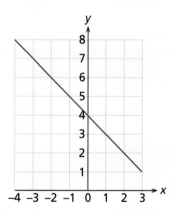

b

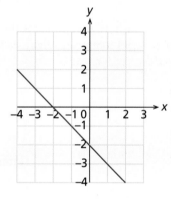

c

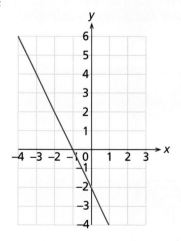

d

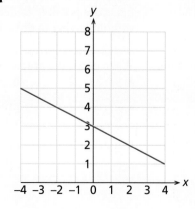

e

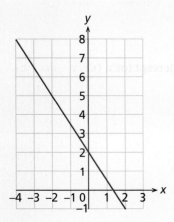

f

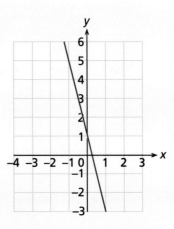

3 a For each of the graphs in questions 1 and 2, calculate the gradient of the straight line.
  b What do you notice about the gradient of each line and its equation?
  c What do you notice about the equation of the straight line and where the line intersects the *y*-axis?

In general, the equation of any straight line can be written in the form:

$$y = mx + c$$

where *m* represents the gradient of the straight line and *c* represents the **intercept** with the *y*-axis (or *y*-intercept). This is shown in the diagram below:

> The intercept is where
> the line crosses the
> *y*-axis.

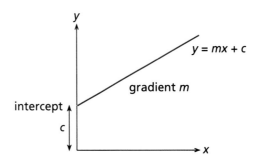

By looking at the equation of a straight line written in the form $y = mx + c$, it is therefore possible to deduce the line's gradient and the intercept with the *y*-axis without having to draw it.

Examples   Write down the gradient and the intercept of the following straight lines:
  a $y = 3x - 2$ **b** $y = -2x + 6$

  a $y = 3x - 2$ gradient $= 3$
                           y-intercept $= -2$

  b $y = -2x + 6$ gradient $= -2$
                           y-intercept $= 6$

Calculate the gradient and the intercept of the following straight lines:

a $2y = 4x + 2$ **b** $y - 2x = -4$ **c** $-4y + 2x = 4$

a $2y = 4x + 2$

This needs to be rearranged into **gradient–intercept** form (i.e. $y = mx + c$).

$$y = \frac{4x + 2}{2}$$

$$y = 2x + 1 \qquad \text{gradient} = 2$$
$$y\text{-intercept} = 1$$

b $y - 2x = -4$

Rearranging into gradient–intercept form:

$$y = 2x - 4 \qquad \text{gradient} = 2$$
$$y\text{-intercept} = -4$$

c $-4y + 2x = 4$

Rearranging into gradient–intercept form:

$$-4y = -2x + 4$$

$$y = \frac{-2x + 4}{-4}$$

$$y = \tfrac{1}{2}x - 1 \qquad \text{gradient} = \tfrac{1}{2}$$
$$y\text{-intercept} = -1$$

Exercise 10.3

For each of the following linear equations, calculate the gradient and the y-intercept.

1 a $y = 2x + 1$ **b** $y = 3x + 5$ **c** $y = x - 2$
    d $y = \tfrac{1}{2}x + 4$ **e** $y = -3x + 6$ **f** $y = -\tfrac{2}{3}x + 1$
    g $y = -x$ **h** $y = -x - 2$ **i** $y = -(2x - 2)$

2 a $y - 3x = 1$ **b** $y + \tfrac{1}{2}x - 2 = 0$ **c** $y + 3 = -2x$
    d $y + 2x + 4 = 0$ **e** $y - \tfrac{1}{4}x - 6 = 0$ **f** $-3x + y = 2$
    g $2 + y = x$ **h** $8x - 6 + y = 0$ **i** $-(3x + 1) + y = 0$

3 a $2y = 4x - 6$ **b** $2y = x + 8$ **c** $\tfrac{1}{2}y = x - 2$
    d $\tfrac{1}{4}y = -2x + 3$ **e** $3y - 6x = 0$ **f** $\tfrac{1}{3}y + x = 1$
    g $6y - 6 = 12x$ **h** $4y - 8 + 2x = 0$ **i** $2y - (4x - 1) = 0$

4 a $2x - y = 4$ **b** $x - y + 6 = 0$ **c** $-2y = 6x + 2$
    d $12 - 3y = 3x$ **e** $5x - \tfrac{1}{2}y = 1$ **f** $-\tfrac{2}{3}y + 1 = 2x$
    g $9x - 2 = -y$ **h** $-3x + 7 = -\tfrac{1}{2}y$ **i** $-(4x - 3) = -2y$

The equation of a line through two points

The equation of a straight line can be deduced once the coordinates of two points on the line are known.

Example   Calculate the equation of the straight line passing through the points $(-3, 3)$ and $(5, 5)$.

Plotting the two points gives:

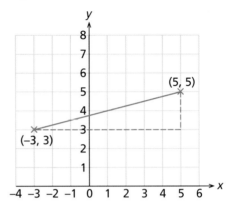

As it is the equation of the straight line that is to be calculated, it takes the general form of $y = mx + c$.

$$\text{gradient} = \frac{5-3}{5-(-3)} = \frac{2}{8} = \frac{1}{4}$$

The equation of the line now takes the form $y = \frac{1}{4}x + c$.

To calculate the value of c, substitute the x- and y-coordinates of one of the points into the equation.

Substituting $(5, 5)$ into the equation gives:

$$5 = \tfrac{1}{4} \times 5 + c$$

$$5 = \tfrac{5}{4} + c$$

Therefore $c = 3\frac{3}{4}$.

The equation of the straight line passing through $(-3, 3)$ and $(5, 5)$ is:

$$y = \tfrac{1}{4}x + 3\tfrac{3}{4}$$

Exercise 10.4

Find the equation of the straight line which passes through each of the following pairs of points.

1 a $(1, 1), (4, 7)$ **b** $(1, 4), (3, 10)$ **c** $(1, 5), (2, 7)$
  d $(0, -4), (3, -1)$ **e** $(1, 6), (2, 10)$ **f** $(0, 4), (1, 3)$
  g $(3, -4), (10, -18)$ **h** $(0, -1), (1, -4)$ **i** $(0, 0), (10, 5)$
2 a $(-5, 3), (2, 4)$ **b** $(-3, -2), (4, 4)$ **c** $(-7, -3), (-1, 6)$
  d $(2, 5), (1, -4)$ **e** $(-3, 4), (5, 0)$ **f** $(6, 4), (-7, 7)$
  g $(-5, 2), (6, 2)$ **h** $(1, -3), (-2, 6)$ **i** $(6, -4), (6, 6)$

SUMMARY

By the time you have completed this chapter you should know:

■ what is meant by the term **gradient**; in mathematics the gradient of a line is found by the formula:

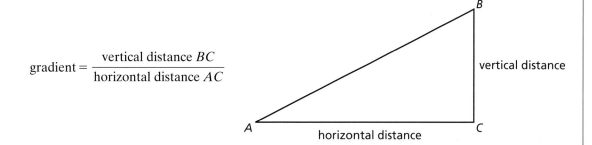

$$\text{gradient} = \frac{\text{vertical distance } BC}{\text{horizontal distance } AC}$$

■ how to calculate the gradient of a straight line, given the coordinates of two points on it, for example if P is at $(3, 2)$ and Q is at $(5, 8)$,

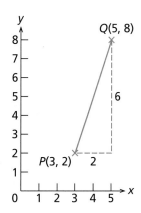

the vertical distance between P and Q is the difference in the y-coordinates, i.e. $8 - 2 = 6$
the horizontal distance between them is the difference in the x-coordinates, i.e. $5 - 3 = 2$
so, by the formula above, the gradient is $\frac{6}{2} = 3$

■ how to rearrange an equation into **gradient–intercept** form, and from that deduce the gradient of the straight line, for example arrange $2y + 4 = 6x$ into the form $y = mx + c$:

$$2y + 4 = 6x$$
$$2y = 6x - 4$$
$$y = 3x - 2$$

so the gradient of the line is 3.

■ how to deduce the equation of a straight line, given the coordinates of two points on it; for example, in the example above where P was at $(3, 2)$ and Q was at $(5, 8)$, the gradient was calculated to be 3. Therefore the equation of that line must be of the form $y = 3x + c$. To calculate the value of c, substitute the x- and y-coordinates of point P into the equation:

$$y = 3x + c$$
$$2 = (3 \times 3) + c$$
$$2 = 9 + c$$
$$-7 = c$$

So the gradient is 3 and the intercept on the y-axis is at -7; $y = 3x - 7$.

Exercise 10A

1 Identify the coordinates of two of the points on the straight line below and use these to find the gradient of the straight line.

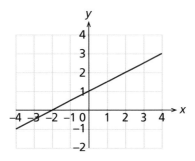

2 For each of the following linear equations, determine the gradient and *y*-intercept.
  a $y = x + 1$　　　　　**b** $y = 3 - 3x$　　　　　**c** $2x - y = -4$　　　　　**d** $2y - 5x = 8$
3 Calculate the gradient of the straight line which passes through each of the following pairs of points.
  a $(1, -1), (4, 8)$　　　　**b** $(0, 7), (3, 1)$

Exercise 10B

1 Identify the coordinates of two of the points on the straight line below and use these to find the gradient of the straight line.

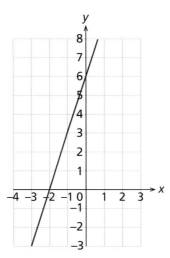

2 For each of the following linear equations, determine the gradient and *y*-intercept.
  a $y = 2x + 3$　　　　　**b** $y = 4 - x$
  c $2x - y = 3$　　　　　**d** $-3x + 2y = 5$
3 Calculate the gradient of the straight line which passes through each of the following pairs of points.
  a $(-2, -9), (5, 5)$　　　　**b** $(1, -1), (-1, 7)$

Exercise 10C

1 The graph on the right shows a set of four parallel lines. Showing your working clearly, calculate the gradient of each of the straight lines.

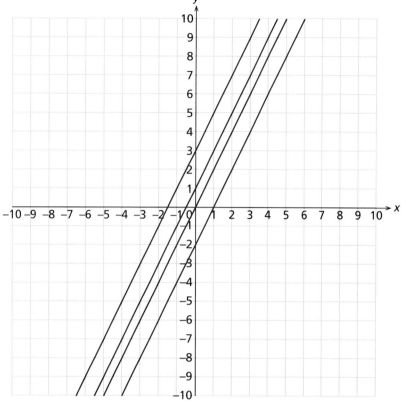

2 The graph on the right shows another set of four parallel lines. For each of the lines determine:
  a its gradient,
  b its intercept with the y-axis,
  c its equation.
  d What conclusions can you make about the gradient and intercept of parallel lines?

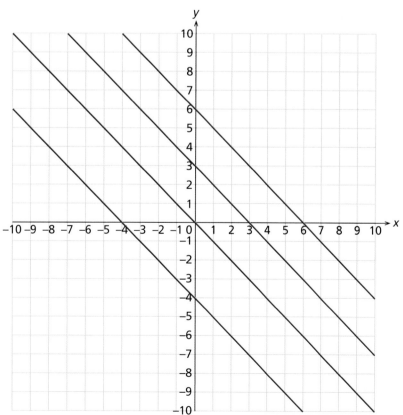

11 Measurement and construction

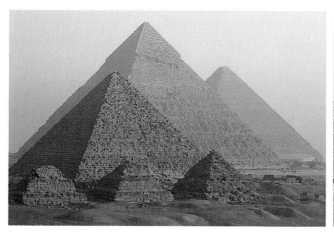

Structures such as the pyramids could only have been planned by engineers with a sound knowledge of two-dimensional figures. The ability to construct squares, triangles and rectangles underpins the methods used by ancient civilisations to construct massive buildings and monuments.

Lines and angles

A straight line can be both drawn and measured accurately using a ruler.

Exercise 11.1

You will need:
- ruler

1 Using a ruler, measure the following lines as accurately as possible and write down how long each line is.

a

b

c

d

e

f

2 Draw lines of the following lengths using a ruler.

   a 3 cm   **b** 8 cm   **c** 4.6 cm   **d** 94 mm   **e** 38 mm   **f** 61 mm

An angle is a measure of turn. When drawn it can be measured using either a protractor or an angle measurer. The units of turn are given in degrees (°).

Examples Measure the angle drawn below.

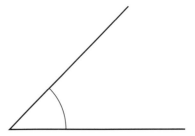

- Place the protractor over the angle so that the cross in the base line lies on the point where the two lines meet.
- Align the 0° with one of the lines drawn.

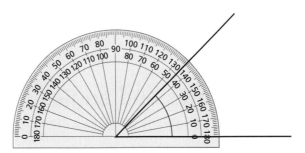

Remember:
The outside numbers on a protractor measure clockwise angles, and the inside numbers measure anti-clockwise angles.

- Decide which scale should be used. In this case the inner scale should be used because it starts at 0° (measuring anti-clockwise).
- Measure the angle using the inner scale.

The angle is 45°.

Draw an angle of 110°.

- Start by drawing a straight line.
- Place the protractor on the line so that the cross is on one of the endpoints of the line. Ensure that the line is aligned with the 0° on the protractor.

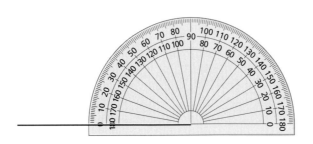

- Decide which scale to use. In this case use the outer scale because it starts at 0° (measuring clockwise).
- Mark where the protractor reads 110°.
- Join the mark made to the endpoint of the line.

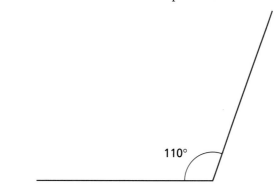

110°

Exercise 11.2

1 Measure each of the following angles.

You will need:
- protractor
- ruler

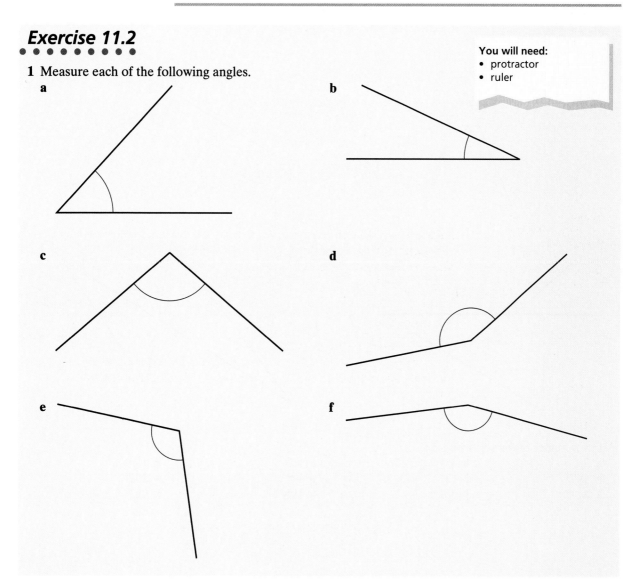

a

b

c

d

e

f

2 Measure each of the angles in the diagrams below.

a

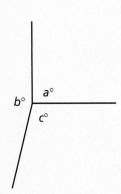

b

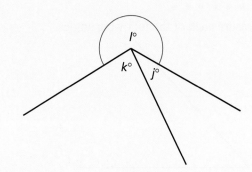

c

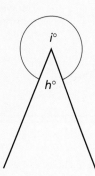

d

e

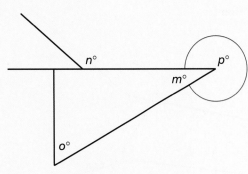

f

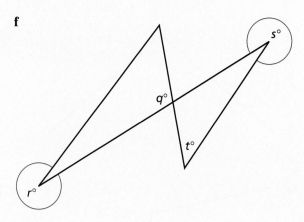

3 Draw angles of the following sizes.

a 20°	**b** 45°	**c** 90°	**d** 120°
e 157°	**f** 172°	**g** 14°	**h** 205°
i 311°	**j** 283°	**k** 198°	**l** 352°

Constructing triangles

Triangles can be drawn accurately by using a ruler and a pair of compasses. This is called **constructing** a triangle.

Example Construct the triangle ABC given that:

$$AB = 8\,cm, BC = 6\,cm \text{ and } AC = 7\,cm$$

- Draw the line AB using a ruler:

A ———————————— 8 cm ———————————— B

> Note that every point on the arc is 6 cm away from B.

- Open up a pair of compasses to 6 cm. Place the compass point on B and draw an arc:

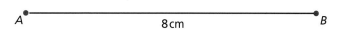

A •———————————— 8 cm ———————————— • B

- Open up the pair of compasses to 7 cm. Place the compass point on A and draw another arc, ensuring it intersects (meets) with the first one. Every point on the second arc is 7 cm from A. Where the two arcs intersect is point C, as it is both 6 cm from B and 7 cm from A.
- Join C to A and C to B:

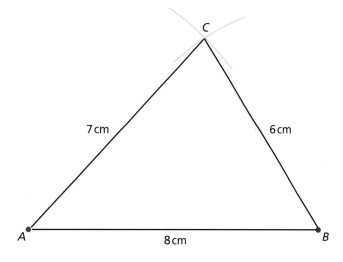

Exercise 11.3

Using only a ruler and a pair of compasses, construct the following triangles.

You will need:
- ruler
- compasses

1 $\triangle ABC$ where $AB = 10\,\text{cm}$, $AC = 7\,\text{cm}$, $BC = 9\,\text{cm}$
2 $\triangle LMN$ where $LM = 4\,\text{cm}$, $LN = 8\,\text{cm}$, $MN = 5\,\text{cm}$
3 $\triangle PQR$, an equilateral triangle of side length $7\,\text{cm}$
4 **a** $\triangle ABC$ where $AB = 8\,\text{cm}$, $AC = 4\,\text{cm}$, $BC = 3\,\text{cm}$
  b Explain why such a triangle is not possible.

Constructing simple geometric figures

'Geometry' means Earth or land measurement, and the Egyptians used it to survey land and buildings.

Exercise 11.4

You will need:
- compasses

1 Draw the following circles using a pair of compasses.

a

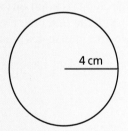

4 cm

b

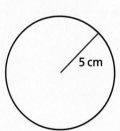

5 cm

c

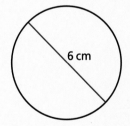

6 cm

2 Using a pair of compasses, copy the following circle patterns.

a

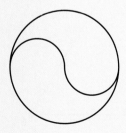

b

3 Draw some circle patterns of your own.

Example	Construct a regular hexagon using a pair of compasses and a ruler.

- Open up a pair of compasses and draw a circle.
- Keeping the compasses open at the same distance, put the point on the circumference of the circle and draw an arc intersecting the circumference.
- Place the compass point on the point of intersection of the arc and the circumference and draw another arc.
- Repeat the above procedure until there are six arcs drawn.

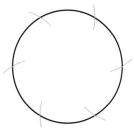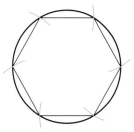

- Draw lines joining the six arcs.

Exercise 11.5

Draw some patterns of your own involving regular hexagons.

Squares and rectangles can also be constructed. To do this, both a ruler and a set square are needed. A set square is used because it provides a right angle with which to draw lines **perpendicular** to each other.

Example	Construct a square of side length 6 cm.

- Draw a line 6 cm long.
- Place a set square at the end of the line, ensuring that one of the perpendicular sides rests on the line drawn.
- Draw a perpendicular line 6 cm long.

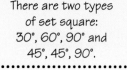
There are two types of set square: 30°, 60°, 90° and 45°, 45°, 90°.

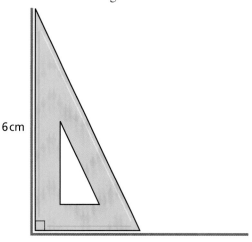

6 cm

6 cm

- Repeat this for the remaining two sides.

Exercise 11.6

Using a ruler and set square only, draw the following shapes.

You will need:
• ruler
• set square
• compasses

1

5 cm

5 cm

2

6 cm

3 cm

3

7 cm

7 cm

7 cm

7 cm

45°

4

5 cm

5 cm

10 cm

Using appropriate geometric equipment, construct the following.

5

6 cm

6 cm

6 cm

6 cm

30°

6

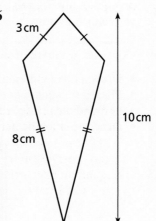

3 cm

8 cm

10 cm

7

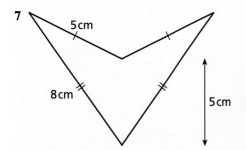

5 cm

8 cm

5 cm

8

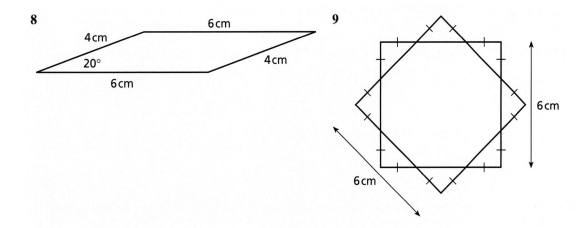

9

Bisecting lines and angles

'Bisect' comes from the Latin *bi*, meaning two, and *sectum*, meaning to cut.

The word **bisect** means to divide or cut something in half. Therefore to bisect an angle means to divide an angle in half. Similarly, to bisect a line means to divide a line in half.

To bisect either a line or an angle involves the use of a pair of compasses.

Examples A line *AB* is drawn below. Construct the perpendicular bisector of *AB*.

Remember:
A perpendicular bisector of a line is a line which divides another one in half and meets it at right angles.

- Open a pair of compasses to more than half the distance *AB*.
- Place the compass point on *A* and draw arcs above and below *AB*.
- With the compasses kept the same distance apart, place the compass point on *B* and draw two more arcs above and below *AB*. Note that the two pairs of arcs should intersect.
- Draw a line through the two points where the arcs intersect.

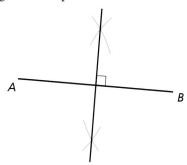

The line drawn is the perpendicular bisector of *AB*, as it divides *AB* in half and also meets it at right angles.

Using a pair of compasses, bisect the angle *A* below.

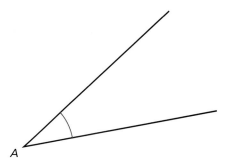

- Open a pair of compasses and place the point on *A*. Draw two arcs so that they intersect the lines (see below).

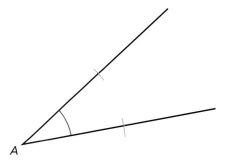

- Place the point of the compasses in turn on the points of intersection and draw another pair of arcs inside the angle. Ensure that they intersect.
- Draw a line through *A* and the point of intersection of the two arcs. This line bisects angle *A* (see below).

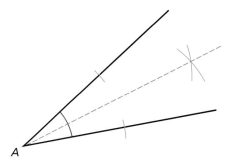

Exercise 11.7

1 Draw each of the lines below on plain paper and construct their perpendicular bisectors.

You will need:
• ruler
• compasses
• protractor

a

A ———————— 8 cm ———————— B

b

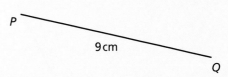

P
9 cm
Q

c

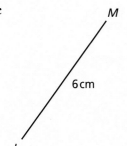

M
6 cm
L

d

C
5 cm
D

2 i) Draw each of the following angles.
ii) Using a pair of compasses, bisect each angle.
  a 45°   **b** 70°   **c** 130°   **d** 173°   **e** 210°   **f** 312°

3 Copy the diagram below.

A
•

B
•

•
C

a Construct the perpendicular bisector of *AB*.
b Construct the perpendicular bisector of *BC*.
c What can be said about the point of intersection of the two perpendicular bisectors?

4 Draw a triangle similar to the one shown below.

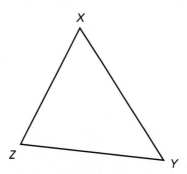

By construction, draw a circle such that points *X*, *Y* and *Z* all lie on its circumference.

5 Draw a triangle similar to the one shown below.

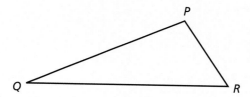

By construction, draw a circle such that points *P*, *Q* and *R* all lie on its circumference.

Scale drawings

Scale drawings are used when an accurate diagram, drawn in proportion, is needed. Common examples of scale drawings are maps and plans. The use of scale drawings involves understanding how to scale measurements.

Examples

A map is drawn to a scale of $1:10\,000$. If two objects are 1 cm apart on the map, how far apart are they in real life? Give your answer in metres.

A scale of $1:10\,000$ means that 1 cm on the map represents 10 000 cm in real life.
Therefore the distance $= 10\,000$ cm
$\qquad\qquad\qquad\qquad = 100$ m

Remember:
Maps and scale drawings always have a scale, usually written as a ratio, for example $1:20\,000$ means that 1 cm on the map is equal to 20 000 cm in real life.

A model boat is built to a scale of $1:50$. If the length of the real boat is 12 m, calculate the length of the model boat in centimetres.

A scale of $1:50$ means that 50 cm on the real boat is 1 cm on the model boat.

$\qquad$ 12 m $= 1200$ cm

Therefore the length of the model boat $= 1200 \div 50$ cm
$\qquad\qquad\qquad\qquad\qquad\qquad\quad = 24$ cm

a Construct triangle ABC such that $AB = 6\,cm$, $AC = 5\,cm$ and $BC = 4\,cm$.
b Measure the perpendicular height of C from AB.
c Calculate the area of the triangle.

a

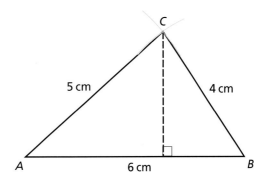

b The perpendicular height is $3.3\,cm$.

c Area $= \dfrac{\text{base length} \times \text{perpendicular height}}{2}$

$\phantom{c \text{ Area}} = \dfrac{6 \times 3.3}{2}$

The area is $9.9\,cm^2$.

Exercise 11.8

You will need:
• ruler
• set square
• compasses

1 In **a–d**, both the scale to which a map is drawn and the distance between two objects on the map are given.
Find the real distance between the two objects, giving your answer in metres.
a $1:10\,000$ $3\,cm$ **b** $1:10\,000$ $2.5\,cm$
c $1:20\,000$ $1.5\,cm$ **d** $1:8000$ $5.2\,cm$

2 In **a–d**, both the scale to which a map is drawn and the real distance between two objects are given.
Find the distance between the two objects on the map, giving your answer in centimetres.
a $1:15\,000$ $1.5\,km$ **b** $1:50\,000$ $4\,km$
c $1:10\,000$ $600\,m$ **d** $1:25\,000$ $1.7\,km$

3 A rectangular pool measures $20\,m$ by $36\,m$ as shown below.

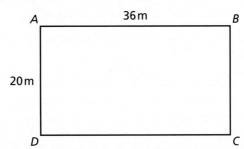

a Construct a scale drawing of the pool, using a scale of $1\,cm$ for every $4\,m$.
b A boy swims across the pool in such a way that his path is the perpendicular bisector of BD. Show, by construction, the path he takes.
c Use your diagram to work out the distance the boy swims.

4 A triangular enclosure is as shown in the diagram below.

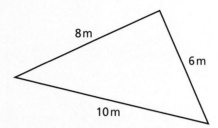

   a Using a scale of 1 cm for each metre, construct a scale drawing of the enclosure.
   b Using your diagram, work out the area of the real enclosure.

5 Three radar stations A, B and C pick up a distress signal from a boat at sea.

  C is 24 km due East of A, AB = 12 km and BC = 18 km. The signal indicates that the boat is **equidistant** (the same distance) from all three radar stations.
   a By construction and using a scale of 1 cm for every 3 km, mark the position of the boat.
   b What is the boat's real distance from the radar stations?

6 A plan view of a field is shown below.

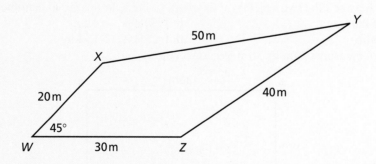

   a Using a scale of 1 cm for every 5 m, construct a scale drawing of the field.
   b A farmer divides the field by running a fence from X in such a way that it bisects $\angle WXY$. By construction show the position of the fence in your diagram.
   c Work out the length of the fencing used.

Bearings

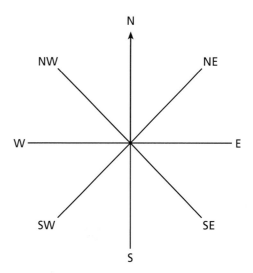

In the days when sailing ships travelled the oceans of the world, compass bearings like the ones in the diagram above were used. To be more accurate, extra points were added to N, S, E, W, NE, SE, SW and NW. Midway between North and North East was North North East, and midway between North East and East was East North East and so on. This gave 32 points to the compass. This was later extended even further to 64 points.

A new system for describing directions was then introduced: the **three-figure bearing** system. North was given the bearing zero. 360° in a clockwise direction was one full rotation.

> **Remember:**
> *Bearings are always measured clockwise from North.*

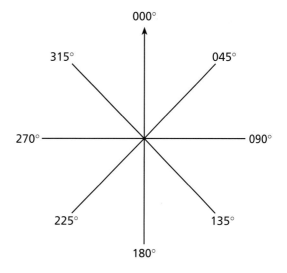

Example   The diagram below shows two ships at positions *A* and *B*.

• *B*

•
A

a What is the bearing of *B* from *A*?
b What is the bearing of *A* from *B*? (This is known as the **back bearing**.)

a To work out a bearing between two objects:

- draw a North arrow at the starting point (i.e. *A*)
- draw a straight line between the two points *A* and *B*
- measure the angle between the North line and the line between the two points, in a clockwise direction
- give your answer as a three-figure bearing (i.e. an angle of 48° would be written as 048°).

The diagram will look as follows:

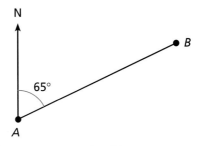

Therefore the bearing of *B* from *A* is 065°.

b The diagram will look as follows:

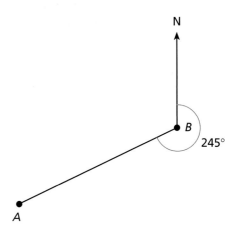

Remember:
Back bearings are ±180° from the original bearing.

The bearing of *A* from *B* is therefore 245°.

Exercise 11.9

Draw diagrams to show the following compass bearings and journeys.
Use a scale of 1 cm : 1 km. North can be taken to be a line vertically up the page.

You will need:
- ruler
- protractor or angle measurer

1 a A boat starts at a point *A*. It travels a distance of 7 km on a bearing
of 135° to point *B*. From *B* it travels 12 km on a bearing of 250° to point *C*.
  b If the boat wishes to make its way back from *C* to *A*, what distance would it travel and on what
  bearing?
  c Another boat wishes to travel directly from *A* to *C*. What is the distance and bearing of this journey?
2 a An athlete starts at a point *P*. He runs a distance of 6.5 km on a bearing of 225° to point *Q*. From *Q*
he runs a further distance of 7.8 km on a bearing of 105° to a point *R*. At *R* he runs towards a point *S*
a further distance of 8.5 km and on a bearing of 090°.
  b Calculate the distance and bearing the athlete has to run to get from *S* back to *P*.
3 a Starting from a point *M*, a horse and rider set off on a bearing of 270° for a distance of 11.2 km to a
point *N*. At *N* they travel 5.8 km on a bearing of 170° to point *O*.
  b What is the bearing and distance of *M* from *O*?
4 In **a–f**, you are given the bearing of point *Y from* point *X*. Draw a diagram to represent the
information and measure the bearing of *X* from *Y*.
  a 130° **b** 145° **c** 220° **d** 200° **e** 152° **f** 234°
5 In **a–d**, you are given the bearing of point *B from* point *A*. Draw a diagram to represent the
information and measure the bearing of *A* from *B*.
  a 300° **b** 320° **c** 290° **d** 282°
6 What conclusions can you make from your answers to questions 4 and 5?

SUMMARY

By the end of this chapter you should know:

- how to draw and measure angles accurately using an angle measurer or protractor (see page 114)
- how to **construct** a triangle using a ruler and a pair of compasses (see page 117)
- how to use a pair of compasses to **bisect** lines and angles (see pages 121 and 122)
- how to carry out calculations involving scale drawings (see page 124)
- how to draw scale diagrams involving **three-figure bearings** (see pages 127 and 128)
- the relationship between a forward and a **back bearing** for a pair of points (see page 128).

Exercise 11A

You will need:
- ruler
- protractor
- compasses
- set square

1 a Using a ruler, measure the length of the line below.

  b Draw a line 4.7 cm long.
2 a Using a protractor, measure the angle below.

  b Draw an angle of 300°.

3 Construct $\triangle ABC$ such that $AB = 8\,$cm, $AC = 6\,$cm and $BC = 12\,$cm.

4 Using a pair of compasses, construct the following circle pattern.

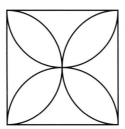

5 Three players P, Q and R are approaching a football. Their positions relative to each other are shown below:

$P \bullet$

$\bullet R$

$Q \bullet$

If the ball is equidistant from all three players, show by construction its position.

6 A plan of a living room is shown below:

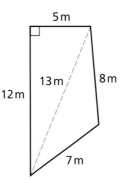

a Using a pair of compasses, construct a scale drawing of the room using $1\,$cm for every metre.
b Using a set square if necessary, calculate the total area of the actual living room.

7 A coastal radar station picks up a distress call from a ship. It is $50\,$km away on a bearing of $345°$. The radar station contacts a lifeboat at sea which is $20\,$km away on a bearing of $220°$.
a Construct a scale drawing of the above situation.
b Use your drawing to find the distance and bearing of the ship from the lifeboat.

Exercise 11B

You will need:
• ruler
• protractor
• compasses

1 Measure each of the five interior angles of the pentagon below:

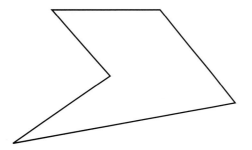

2 Using a ruler and a protractor, draw a triangle with angles of 40°, 60° and 80°.

3 a Draw an angle of 320°.

  b Using a pair of compasses, bisect the angle.

4 Construct the circle pattern shown below.

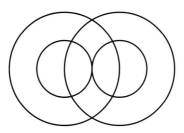

5 In **a** and **b**, both the scale to which a map is drawn and the real distance between two objects are given. Find the distance between the two objects on the map, giving your answer in centimetres.

  a 1:20 000   4.4 km   **b** 1:50 000   12.2 km

6 a Construct a regular hexagon with side length 3 cm.

  b By showing your method clearly, calculate its area.

7 An aircraft is seen on radar at Milan airport. The aircraft is 210 km away from the airport on a bearing of 065°. The aircraft is diverted to another airport, which is 130 km away from Milan on a bearing of 215°. Use an appropriate scale drawing to find the distance and bearing of the other airport from the aircraft.

Exercise 11C

Two points X and Y lie on a straight line as shown below. A pair of compasses has been used to demonstrate that four points A, B, C and D, each of which is equidistant from X and Y, lie on a straight line known as the perpendicular bisector.

You will need:
- ruler
- compasses

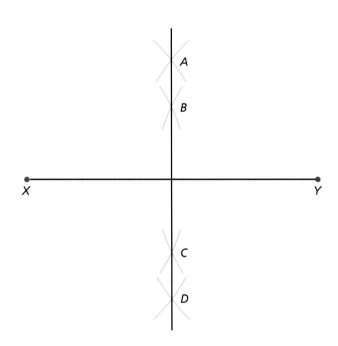

- Using a pair of compasses, investigate the line made by points which are twice the distance from X as they are from Y (i.e. the ratio of distances is $2:1$).
- Using a pair of compasses, investigate the line made by points which are three times the distance from X that they are from Y (i.e. the ratio of distances is $3:1$).
- Investigate the change in shape of the line made by points where the ratio of distances from X and Y is $n:1$.

Exercise 11D

You will need:
• computer with Cabri II or similar software installed

• Using Cabri II, or a similar geometry package, demonstrate that it is possible, by geometric construction, to draw a circle through the three vertices of a triangle. One example is shown in the diagram below.

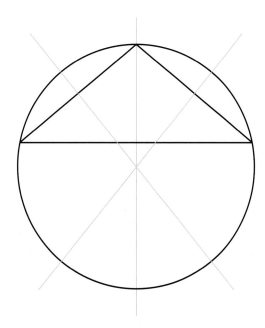

• Explain, using appropriate mathematical language, how you constructed your diagram.

Exercise 11E

On 23 February 532 AD, the Emperor Justinian laid the foundation stone for the Church of the Holy Wisdom, or Hagia Sophia, in Constantinople (modern Istanbul). Most of the population of Istanbul is Muslim, and Hagia Sophia is a museum. After Constantinople was taken over by Sultan Mohammed II, he turned Hagia Sophia into a mosque, later adding minarets.

This building is considered a masterpiece of construction. Try to obtain pictures of it, and study them from a mathematical perspective.

Which English cathedral was greatly influenced by Hagia Sophia?

12 Geometrical terms and relationships

Spain was conquered in 714 AD by Muslim Moorish armies. During the 800 years until Spain was re-conquered by Christians, the Moors greatly influenced the culture of Spain.

The Alhambra is a walled city and fortress in Granada, Spain. It was built during the last Islamic sultanate on the Iberian Peninsula. The palace is lavishly decorated with stone and wood carvings, and tile patterns on most of the ceilings, walls and floors. In Islamic art no representations of living beings are permitted, so geometric patterns, especially symmetrical ones, are used.

The picture below is an example of one of the tilings at the Alhambra and demonstrates how an understanding of angle can lead to the creation of a beautiful symmetrical pattern.

> The word 'angle' comes from the Latin *angulus*, meaning corner.

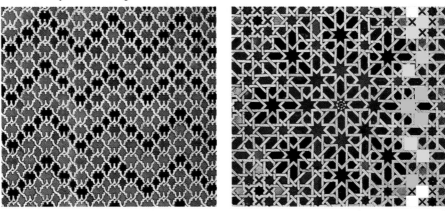

An angle is a measure of turn. People often say sentences such as 'turn round the corner' or 'turn over a page'; these both represent an amount of turn and, as a result, an angle.

The most common unit of measurement for an angle is the **degree** (°). In this chapter we will look mainly at the special angle properties and relationships that exist.

> **Remember:**
> *A full turn is 360°.*
> *A half-turn is 180°.*
> *A quarter-turn is 90°.*

Angles on a straight line

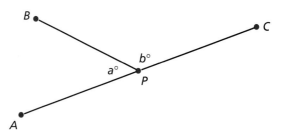

The points *APC* lie on a straight line. A person standing at point *P*, initially facing point *A*, turns through an angle $a°$ to face point *B* and then turns a further angle $b°$ to face point *C*. The person has turned through half a complete turn and therefore rotated through 180°. Therefore $a° + b° = 180°$. This can be summarised as follows.

● *Angles on a straight line, about a point, add up to 180°.*

Example

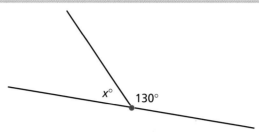

Calculate the value of *x*.

$$x° + 130° = 180°$$
$$x° = 180° - 130°$$

Therefore *x* = 50.

Angles at a point

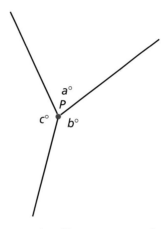

In the diagram above it can be seen that if a person standing at *P* turns through each of the angles *a*°, *b*° and *c*° in turn, then the total amount he has rotated would be 360° (a complete turn). Therefore

$$a° + b° + c° = 360°$$

● *Angles about a point add up to 360°.*

Exercise 12.1

1 Calculate the size of *x* in each of the following.

a

b

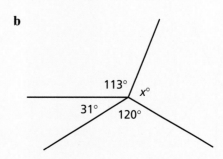

c

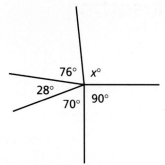

d

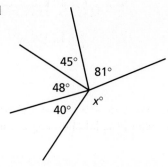

2 In each of the following diagrams, the angles lie about a point on a straight line. Calculate the size of *y* in each case.

a

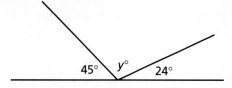

b

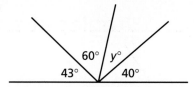

c

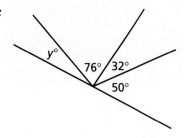

d

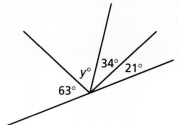

3 Calculate the size of *p* in each of the following.

a

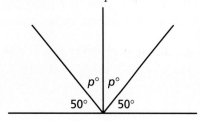

b

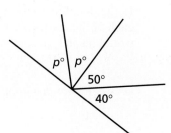

c

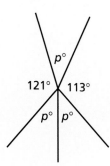

d

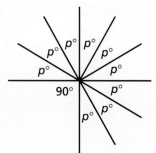

Angles formed with intersecting and parallel lines

Exercise 12.2

You will need:
- ruler
- protractor

1 Draw a similar diagram to the one shown below.
Measure carefully each of the labelled angles and write them down.

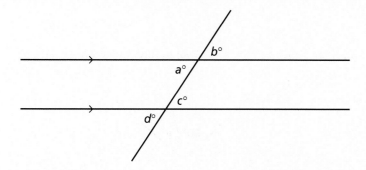

Remember:
The outside numbers on a protractor measure clockwise angles, and the inside numbers measure anti-clockwise angles.

2 Draw a similar diagram to the one shown below. Measure carefully each of the labelled angles and write them down.

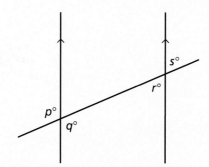

3 Draw a similar diagram to the one shown below. Measure carefully each of the labelled angles and write them down.

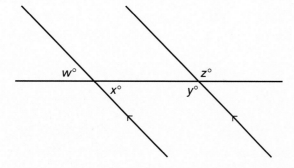

4 Write down what you have noticed about the angles you measured in questions 1–3.

Remember:
Vertically opposite angles can be found by looking for an 'X' formation in a diagram.

When two straight lines cross, we find that the angles opposite each other are the same size. They are known as **vertically opposite angles**. By using the fact that angles at a point on a straight line add up to 180°, we can show why vertically opposite angles must always be equal in size.

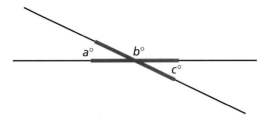

$$a° + b° = 180°$$
$$c° + b° = 180°$$

Therefore, a is equal to c.

Exercise 12.3

You will need:
• ruler
• protractor

1 Draw a similar diagram to the one shown below. Measure carefully each of the labelled angles and write them down.

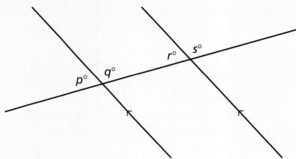

2 Draw a similar diagram to the one shown below. Measure carefully each of the labelled angles and write them down.

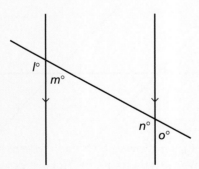

3 Draw a similar diagram to the one shown below. Measure carefully each of the labelled angles and write them down.

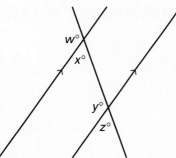

4 Write down what you have noticed about the angles you measured in questions 1–3.

> **Remember:**
> *Corresponding angles can be found by looking for an 'F' formation in a diagram.*
> *Sometimes the 'F' may appear back to front or upside down.*

When a line intersects two parallel lines, as in the diagrams below, we find that certain angles are the same size.

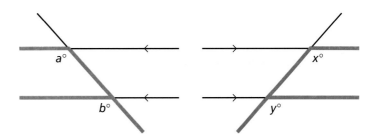

The angles *a* and *b* are equal and are known as **corresponding** angles. Similarly, angles *x* and *y* are equal.

A line intersecting two parallel lines also produces another pair of equal angles known as **alternate** angles. These can be shown to be equal by using the fact that both vertically opposite and corresponding angles are equal.

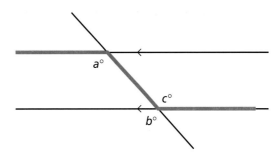

> **Remember:**
> *Alternate angles can be found by looking for a 'Z' formation in a diagram.*

In the diagram above, $a° = b°$ (corresponding angles). But $b° = c°$ (vertically opposite). It can therefore be deduced that $a° = c°$.

Angles *a* and *c* are alternate angles.

Exercise 12.4

In each of the following questions, some of the angles are given. Deduce, giving your reasons, the size of the unknown labelled angles.

1

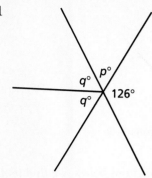

2

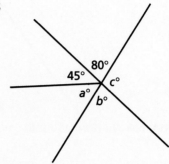

3

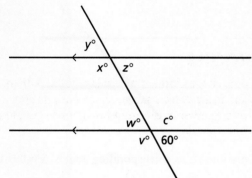

4

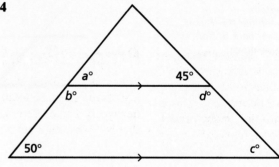

5

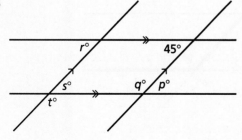

6

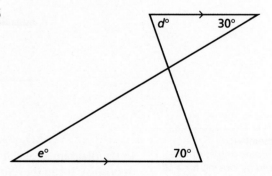

7

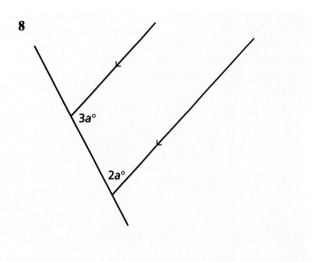

8

Properties of polygons

The word 'polygon' comes from the Greek *poly*, meaning 'many', and *gonia*, meaning 'angle'.

Polygons are closed, two-dimensional shapes formed by straight lines, with at least three sides. Examples of polygons include triangles, quadrilaterals and hexagons. The pattern below shows a number of different polygons **tessellating**; that is, fitting together with no gaps or overlaps.

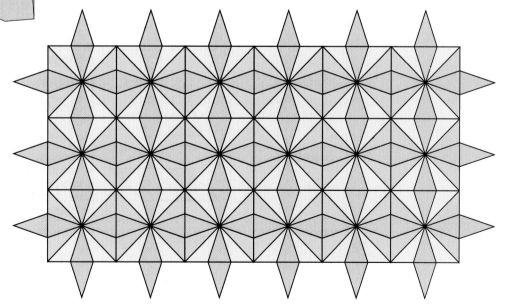

The polygons in the pattern on page 141 tessellate because they have particular properties. The properties of several triangles and quadrilaterals are listed below:

name	shape	properties
isosceles triangle		• two of its sides are the same length • two angles are the same size
equilateral triangle		• all sides are the same length • all angles are the same size **Note.** An equilateral triangle is a special isosceles triangle.
scalene triangle		• all sides are a different length • all angles are a different size
square		• all sides are of equal length • all interior angles are right angles • diagonals intersect at right angles
rectangle		• opposite sides are of equal length • all interior angles are right angles **Note.** A square is a special rectangle.
rhombus		• all sides are of equal length • two pairs of parallel sides • opposite angles are equal • diagonals intersect at right angles
parallelogram		• opposite sides are equal • opposite angles are equal • two pairs of parallel sides **Note.** A rhombus is a special parallelogram.
trapezium		• one pair of parallel sides
kite		• two pairs of equal sides • one pair of equal angles • diagonals intersect at right angles

Exercise 12.5

1 By looking at the tessellating pattern on page 141, identify as many different polygons as you can.
2 Name each of the polygons drawn below. Give reasons for your answer.

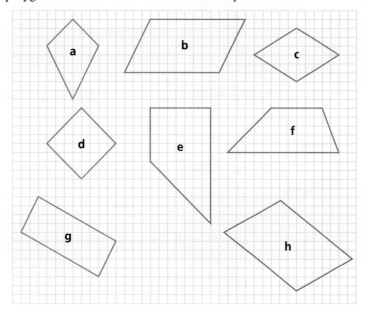

Angle properties of polygons

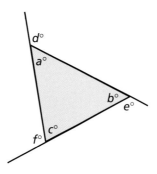

In the triangle above, the **interior angles** are labelled $a°$, $b°$ and $c°$, and the **exterior angles** are labelled $d°$, $e°$ and $f°$.

Imagine a person standing at one of the **vertices** of the triangle and walking along the edges of the triangle until they are at the start again. At each vertex they would have turned through an angle equivalent to the exterior angle at that point. Therefore, during the complete journey, they would have turned through an angle equivalent to one complete turn, i.e. 360°:

$$d° + e° + f° = 360°$$

It is also true that $a° + d° = 180°$ (angles on a straight line). Similarly $b° + e° = 180°$ and $c° + f° = 180°$. Therefore

$$a° + b° + c° + d° + e° + f° = 540°$$
$$a° + b° + c° + 360° = 540°$$
$$a° + b° + c° = 180°$$

These findings lead us to two more important rules.

● *The exterior angles of a triangle (indeed of any polygon) add up to 360°.*
● *The interior angles of a triangle add up to 180°.*

By looking once again at the triangle, it can now be stated that:

$$a° + d° = 180°$$

and also

$$a° + b° + c° = 180°$$

Therefore

$$d° = b° + c°$$

So the exterior angle of a triangle is equal to the sum of the opposite two interior angles.

Exercise 12.6

1 For each of these triangles, use the information given to calculate the size of x.

a

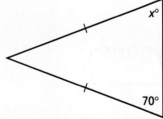

b

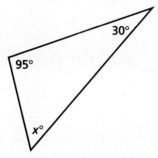

c

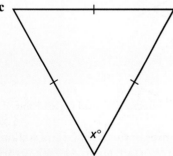

d

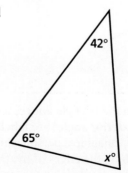

e

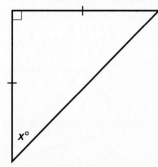

f

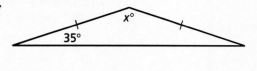

2 For each of the diagrams below, calculate the size of the unknown labelled angles.

a

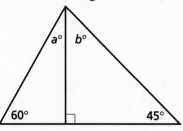

b

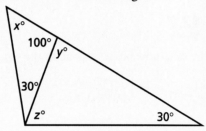

c

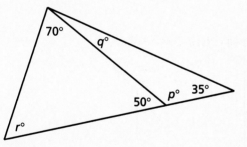

d

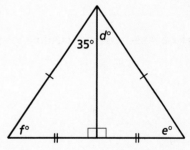

e

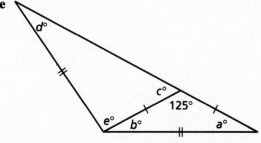

f

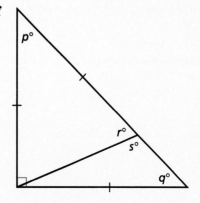

In each of the quadrilaterals below, a straight line is drawn from one of the vertices to the opposite vertex. The result is to split the quadrilaterals into two triangles.

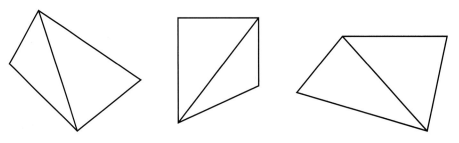

Therefore, as a quadrilateral can be split into two triangles, the sum of the four angles of any quadrilateral must be 360°.

Exercise 12.7

For each of the diagrams below, calculate the size of the unknown labelled angles.

1

$100°$

$a°$

$75°$ $70°$

2

$y°$

$x°$

$40°$

$z°$

3

$m°$

$85°$ $125°$

$n°$

4

$65°$

$u°$

$t°$

$s°$

5

$j°$ $i°$

$k°$ $h°$ / $60°$

6

$e°$ $c°$ $a°$

$d°$

$b°$ $80°$

7

$45°$

$p°$

$50°$ $q°$ $r°$

8

$r°$

$v°$

$50°$

$p°$

$u°$ $t°$ $s°$ $q°$ $75°$

As stated earlier, a polygon is a closed two-dimensional shape, bounded by straight lines. A **regular polygon** is distinctive in that all its sides are of equal length and all its angles of equal size. Below are some examples of regular polygons.

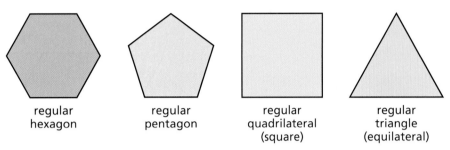

| regular hexagon | regular pentagon | regular quadrilateral (square) | regular triangle (equilateral) |

The name of each polygon is derived from the number of angles it contains. The following list identifies some of these polygons.

  3 angles = **tri**angle
  4 angles = **quad**rilateral (tetragon)
  5 angles = **pent**agon
  6 angles = **hex**agon
  7 angles = **hept**agon (septagon)
  8 angles = **oct**agon
  9 angles = **non**agon
  10 angles = **dec**agon
  12 angles = **dodec**agon

The sum of the interior angles of a polygon

In the polygons below, a straight line is drawn from each vertex to vertex A.

> These shapes are **irregular** polygons since their sides are not of equal length.

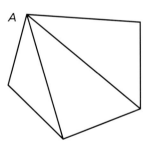
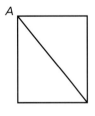
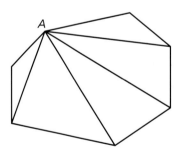

As can be seen, the number of triangles is always 2 less than the number of sides the polygon has, i.e. if there are n sides, then there will be $(n-2)$ triangles. Since the angles of a triangle add up to $180°$, the sum of the interior angles of a polygon is therefore $180(n-2)$ degrees.

Example Find the sum of the interior angles of a regular pentagon and hence the size of each interior angle.

For a pentagon, $n = 5$.
Therefore the sum of the interior angles
$$= 180(5-2)°$$
$$= (180 \times 3)°$$
$$= 540°$$
For a regular pentagon the interior angles are of equal size.
Therefore each angle is $540 \div 5 = 108°$.

The sum of the exterior angles of a polygon

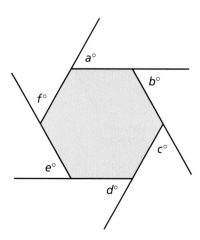

The angles marked $a°$, $b°$, $c°$, $d°$, $e°$ and $f°$ represent the exterior angles of the regular hexagon. As we have already found, the sum of the exterior angles of any polygon is 360°. So, in a regular hexagon each exterior angle is 60°.

If the polygon is regular and has n sides, then

each exterior angle is $\dfrac{360°}{n}$.

Examples   Find the size of an exterior angle of a regular nonagon.

A nonagon has nine sides, so the exterior angle is

$$\frac{360°}{9} = 40°$$

Calculate the number of sides a regular polygon has if each exterior angle is 15°.

Let n be the number of sides.

$$\frac{360°}{n} = 15°$$

so

$$360° = 15° \times n$$

$$n = \frac{360}{15}$$

$$= 24$$

The polygon has 24 sides.

Exercise 10D

You will need:
• computer with a graphing package installed
    or
• graphical calculator

Using a graphing package or graphical calculator, try to draw straight lines so that your screen resembles the diagram below.

Explain clearly how you arrived at the equation for each straight line.

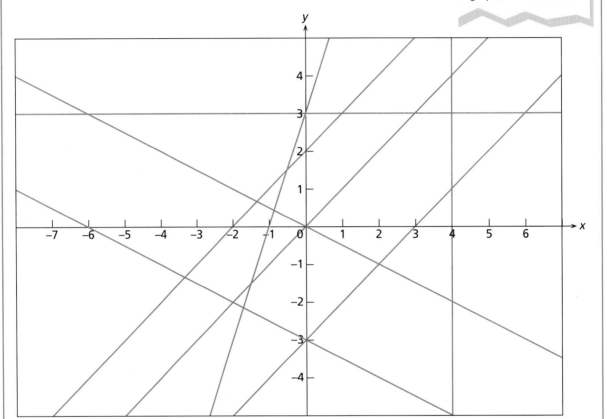

Exercise 10E

Gradients of railway lines are not measured using the formula

$$\text{gradient} = \frac{\text{vertical height}}{\text{horizontal distance}}$$

Find out how railway gradients are measured.

Exercise 12.8

1 Find the sum of the interior angles of the following polygons.
   a a hexagon   **b** a nonagon   **c** a heptagon

2 Find the size of each interior angle of the following regular polygons.
   a an octagon   **b** a square   **c** a decagon
   d a dodecagon

3 Find the size of each exterior angle of the following regular polygons.
   a a pentagon   **b** a dodecagon   **c** a heptagon

4 The sizes of the exterior angle of regular polygons are given below. In each case calculate the number of sides the polygon has.
   a 20°   **b** 36°   **c** 10°
   d 45°   **e** 18°   **f** 3°

5 The sizes of the interior angle of regular polygons are given below. In each case calculate the number of sides the polygon has.
   a 108°   **b** 150°   **c** 162°
   d 156°   **e** 171°   **f** 179°

6 Calculate the number of sides a regular polygon has if an interior angle is five times the size of an exterior angle.

7 Copy and complete the table below for regular polygons.

number of sides	name	sum of exterior angles	size of an exterior angle	sum of interior angles	size of an interior angle
3					
4					
5					
6					
7					
8					
9					
10					
12					

SUMMARY

By the end of this chapter you should know:

■ that the angles on a straight line, about a point, add up to 180°

$$a° + b° = 180°$$

■ that the angles about a point add up to 360°

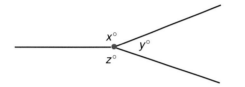

$$x° + y° + z° = 360°$$

■ how to identify **vertically opposite**, **corresponding** and **alternate** angles

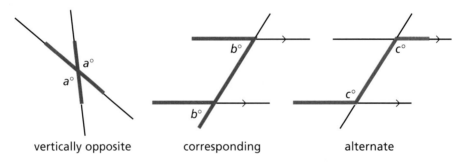

vertically opposite   corresponding   alternate

■ what is meant by the term **tessellating**
■ the geometric properties of common polygons
■ how to calculate the **interior** and **exterior angles** of **regular polygons**.

Exercise 12A

1 For each of these diagrams, calculate the size of the unknown labelled angles.

a

b

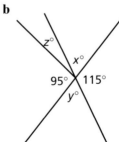

c
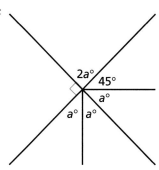

d
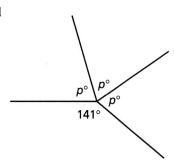

2 For each of the diagrams below, calculate the size of the unknown labelled angles.

a
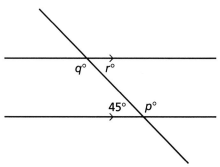

b
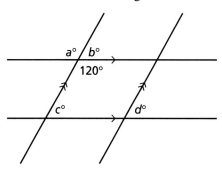

3 For the diagrams below, calculate the size of the unknown labelled angles.

a
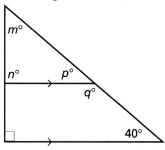

b
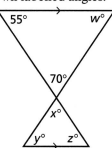

c
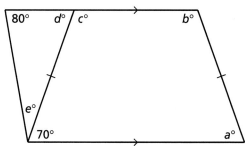

d
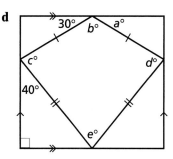

4 Draw a diagram of an octagon to help illustrate the fact that the sum of the internal angles of an octagon is given by
$$180 \times (8 - 2)°$$

You will need:
• ruler
• protractor

5 Find the size of each interior angle of a 20-sided regular polygon.
6 What is the sum of the interior angles of a nonagon?
7 What is the size of the exterior angle of a regular pentagon?
8 Explain why an equilateral triangle is a special isosceles triangle.

Exercise 12B

1 For each of the diagrams below, calculate the size of the unknown labelled angles.

a

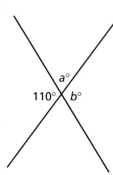

b

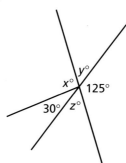

c

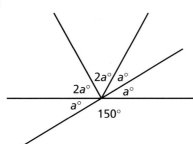

d

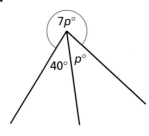

2 For each of the diagrams below, calculate the size of the unknown labelled angles.

a

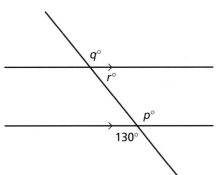

b

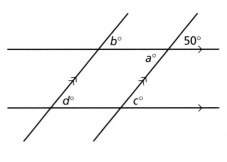

3 For each of the diagrams below, calculate the size of the unknown labelled angles.

a

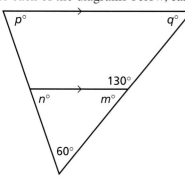

b

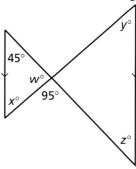

c

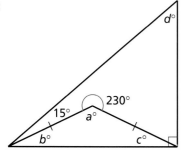

d

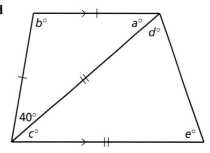

4 Draw a diagram of a hexagon to help illustrate the fact that the sum of the internal angles of a hexagon is given by
$$180 \times (6 - 2)^\circ$$

You will need:
• ruler
• protractor

5 Find the value of each interior angle of a regular polygon with 24 sides.
6 What is the sum of the interior angles of a regular dodecagon?
7 What is the size of the exterior angle of a regular dodecagon?
8 Explain why a square is a special kind of rhombus.

Exercise 12C Ma1

You will need:
• ruler
• protractor

As you have seen throughout the chapter, regular polygons have particular angle properties.

• Investigate (practically or otherwise), giving reasons, which of the above regular polygons tessellate by themselves and which tessellate with other polygons.
• Investigate which other types of polygons tessellate.
• Draw a tessellating pattern to demonstrate your findings.

Exercise 12D

Using Cabri II or a similar geometrical package, demonstrate each of the rules covered in this chapter. The example below shows how Cabri II can be used to show that angles around a point add up to 360°.

You will need:
- computer with Cabri II or similar software installed

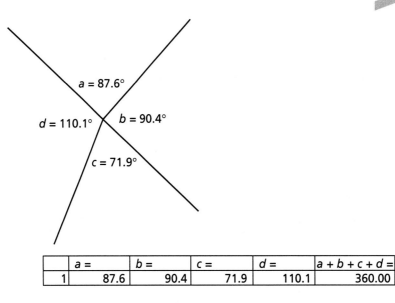

	a =	b =	c =	d =	a + b + c + d =
1	87.6	90.4	71.9	110.1	360.00

Exercise 12E

Find out more about the Alhambra in Granada, Spain. You could answer the following questions:

1 Who built the Alhambra and when?
2 How long were the Moors in Spain?
3 What other great Moorish cities are there in Europe?
4 What areas of mathematics and art were brought to Europe by the Moors?
5 Who was Othello?

13 The circle

We have the ability to visualise solid (three-dimensional) shapes, but often find it easier to think in flat (two-dimensional) shapes. Most people think of the illustrations below as examples of circles. You can investigate most of them further using the internet if you wish.

Vocabulary of the circle

The circle has specific vocabulary to identify different parts of it. The diagram on the right identifies some of the main parts of the circle and gives their names.

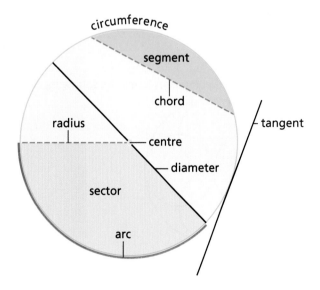

Exercise 13.1

The diagram of the circle on page 155 includes terms which are defined below. Copy and complete the sentences.

1 A line that is always the same distance from a single fixed point is called a _____ .
2 The perimeter of a circle is its _____ .
3 A straight line from the centre of a circle to the circumference of the circle is a _____ (plural _____).
4 A straight line across a circle, which starts and ends at two points on the circumference, is a _____ .
5 A chord which passes through the centre of a circle is called a _____ .
6 A line which forms part of the circumference of a circle is called an _____ .
7 The area enclosed by two radii and an arc is called a _____ .
8 The area enclosed by an arc and a chord is called a _____ .
9 A straight line which just touches the circumference of a circle is called a _____ .

Constructing simple geometric figures starting from circles

To construct circles accurately, a pair of compasses should be used. However, to draw a circle properly using a pair of compasses requires practice.

Exercise 13.2

You will need:
• pair of compasses

1 Draw the following circles using a pair of compasses. In each case O is the centre.

a

3 cm

b

2.5 cm

c
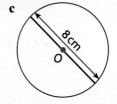
8 cm

2 Using a pair of compasses, copy the following circle patterns.

a

b

3 Draw some circle patterns of your own.

4 Construct a regular hexagon using a pair of compasses and a ruler.

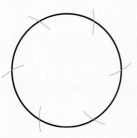 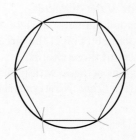

> **Remember:**
> Chapter 11 Measurement and construction
> Draw your construction lines faintly.

> **Remember:**
> Perpendicular bisectors.
>
>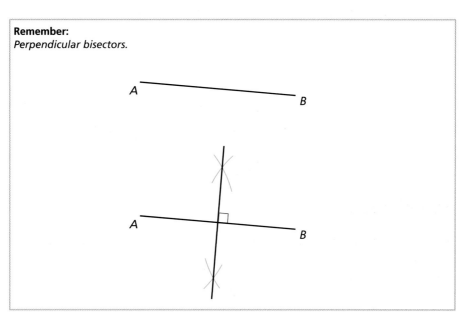

Exercise 13.3

1 Copy each of the lines drawn below on plain paper and construct their perpendicular bisectors.

> **You will need:**
> • pair of compasses
> • ruler

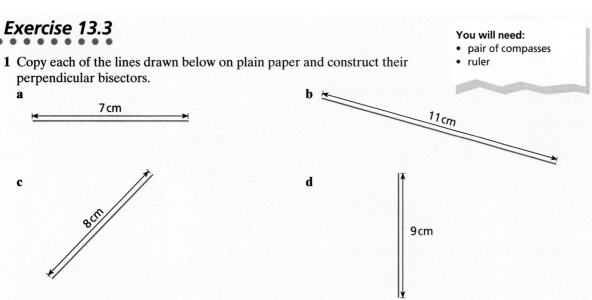

a

7 cm

b

11 cm

c

8 cm

d

9 cm

2 a Draw a circle of approximate radius 6 cm.
  b Draw a diameter of the circle.
  c Construct the perpendicular bisector of the diameter.
  d Use your construction to draw a second diameter at right angles to the first.
  e Draw four chords from where the diameters meet the circumference of the circle.
  f Construct the perpendicular bisectors of these chords.
  g Use the constructions to draw a regular octagon within the circle.
3 Draw a circle of diameter approximately 12 cm. Now construct a regular octagon inside this circle.
4 Copy the diagram below.

P •

Q •

• R

  a Construct the perpendicular bisector of *PQ*.
  b Construct the perpendicular bisector of *QR*.
  c What can be said about the point of intersection of the two perpendicular bisectors?
5 Draw a triangle similar to the one shown below.

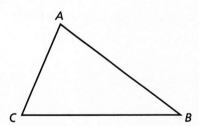

By construction, draw a circle to pass through points *A*, *B* and *C*. This is called the **circumcircle** of the triangle.
6 Draw a triangle similar to the one shown below.

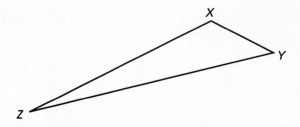

By construction, draw a circle to pass through points *X*, *Y* and *Z*.

SUMMARY

By the time you have completed this chapter you should know:

■ the vocabulary of the circle (see page 155)
■ how to construct a regular hexagon and regular octagon within a circle (see page 157)
■ how to construct the **circumcircle** of a triangle (see Exercise 13.3 question 5).

Exercise 13A

You will need:
• pair of compasses
• ruler

1 Draw a circle and mark on it:
   a a radius **b** an arc **c** a segment
   d a tangent to the circle
2 Draw a line approximately 15 cm long and construct its perpendicular bisector.
3 Draw three points P, Q, R, not in a straight line. By construction find and mark a point equidistant (the same distance) from P, Q and R.
4 Construct an equilateral triangle inside a circle.

Exercise 13B

You will need:
• pair of compasses
• ruler

1 Define the following terms.
   a a diameter of a circle
   b a sector of a circle
   c the circumference of a circle
2 Draw a triangle LMN and then construct its circumcircle.
3 Three lionesses L_1, L_2 and L_3 are positioned as shown, equidistant from a gazelle.

$L_1 \bullet$

$\bullet L_2$

$\bullet L_3$

Draw a diagram similar to the one above and, by construction, find the position of the gazelle.
4 Using a pair of compasses, construct a square inside a circle.

Exercise 13C

You will need:
• pair of compasses

A computer-aided design company called OCC commissions a new logo.
By construction, produce two designs from which the directors can choose.

Exercise 13D

You will need:
• computer with LOGO installed

Using the software package LOGO, design circle patterns of your choice. Some examples are given below.

Exercise 13E

Karl Frederick Gauss is often considered to be the greatest mathematician of all time. He thought his finest discovery was a method of constructing a certain regular polygon. How many sides did this polygon have?

14 Transformations

Tiger! Tiger! burning bright
In the forests of the night,
What immortal hand or eye
Could frame thy fearful symmetry?

William Blake (1757–1827) was an engraver, a painter and a poet. Although during his lifetime he wasn't well known as a poet, *The Tiger* has since become one of the most well-known poems in the English language.

An **object** undergoing a **transformation** changes in either position or shape. In its simplest form, this change can occur as a result of either a **reflection** or a **rotation**. With either of these transformations though, only the position of the object changes. If an object undergoes a transformation, then its new position is known as the **image**. Because the shape of the object remains unchanged, the object and image are said to be **congruent**.

> **Remember:**
> If an object and an image are congruent, each will fit exactly over the other.

Reflection

> **Remember:**
> The mirror line is the *line of symmetry*.

You will need:
- squared paper
- ruler

for most of the exercises in this chapter.

If an object is reflected it undergoes a 'flip' movement about a dotted line known as the **mirror line**, for example

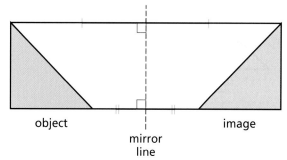

object   image

mirror
line

A point on the object and its equivalent point on the image are **equidistant** from the mirror line, this distance being measured at right angles to the mirror line.

Exercise 14.1

Copy each of the following diagrams and draw in the position of the image(s).

1

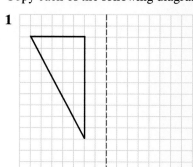

2

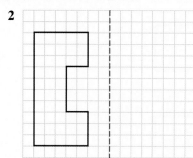

3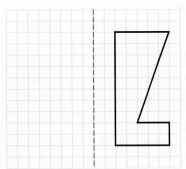

4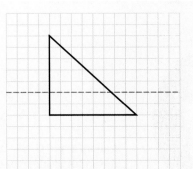

5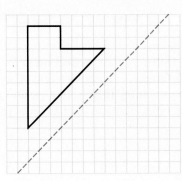

6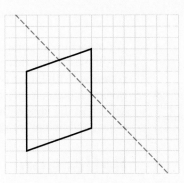

7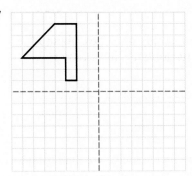

8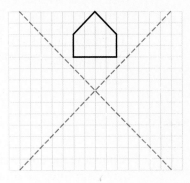

Exercise 14.2

Copy the following objects and images and in each case draw in the position of the mirror line(s).

1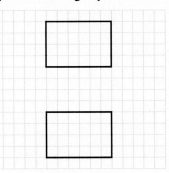

2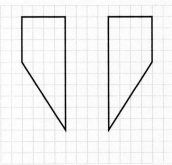

3

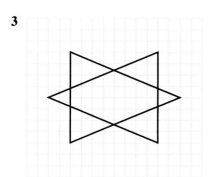

4

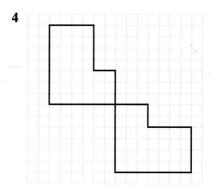

5

6

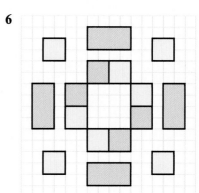

It is essential to know the position of the mirror line when describing a reflection. At times, its equation as well as its position will be required.

Examples Find the equation of the mirror line in the diagram below.

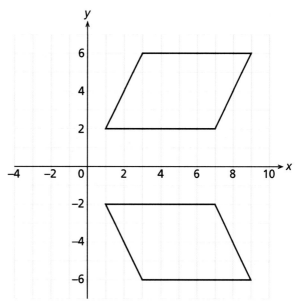

Here the mirror line is the *x*-axis. The equation of the mirror line is therefore $y = 0$.

Copy the diagram below, then:

a draw in the position of the mirror line,
b give the equation of the mirror line.

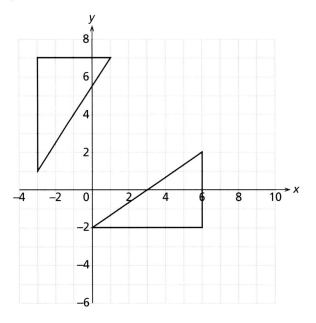

a

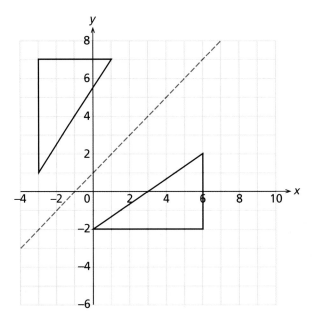

b The equation of the mirror line is $y = x + 1$.

Exercise 14.3

Copy each of the following diagrams, then:
a draw the position of the mirror line(s),
b give the equation of the mirror line(s).

1

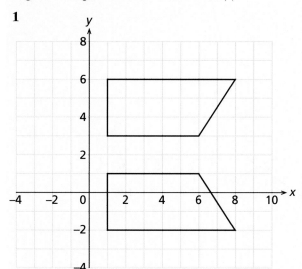

2

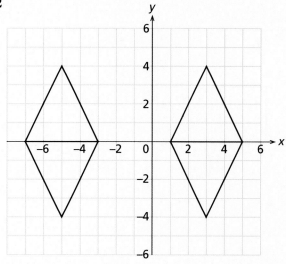

3

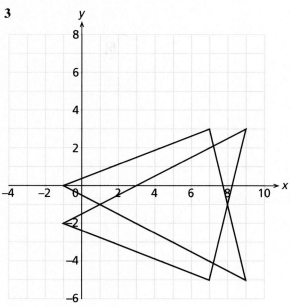

4

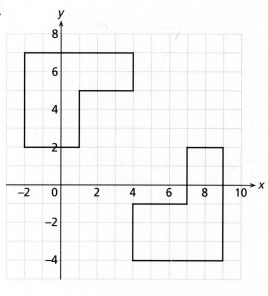

5

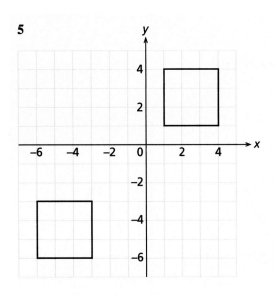

6

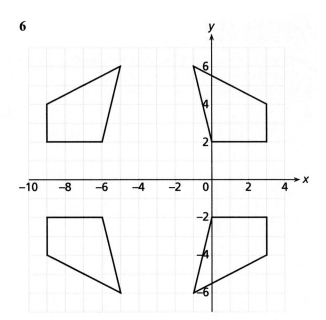

7

8

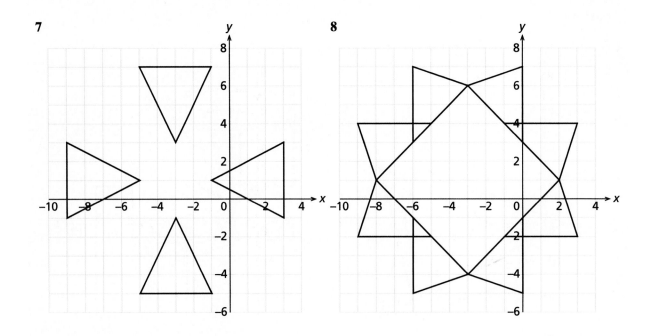

Exercise 14.4

For the following questions, draw a new grid each time and reflect the object in each of the lines given. Make sure your grid is big enough!

1

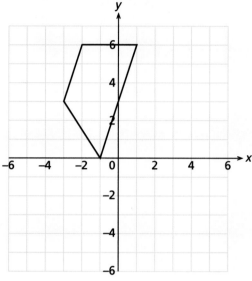

a $x = 2$
b $y = 0$
c $y = x$
d $y = -x$

2

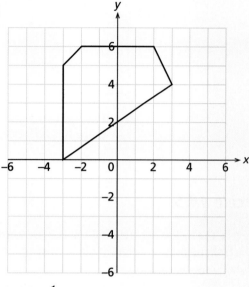

a $x = -1$
b $y = -x - 1$
c $y = x + 2$
d $x = 0$

3 Reflect the triangles in the following pairs of lines:

$x = 1$ and $y = -3$

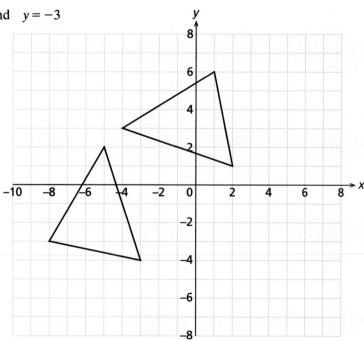

Reflection and three-dimensional objects

A **plane of symmetry** divides a three-dimensional (solid) shape into two congruent solid shapes, for example:

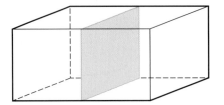 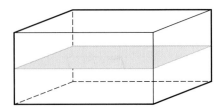

A cuboid has two planes of symmetry as shown.

A shape has **reflective symmetry** if it has one or more lines or planes of symmetry.

Exercise 14.5

For each of the following questions, make two copies of the solid shape shown, then:

a on each shape draw a different plane of symmetry,

b calculate how many planes of symmetry the shape has in total.

1

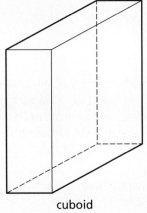

cuboid

2

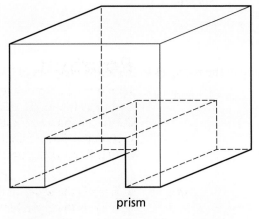

prism

3

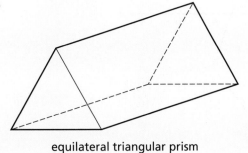

equilateral triangular prism

4

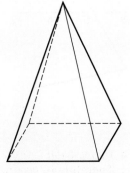

square-based pyramid

5

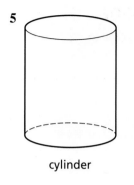

cylinder

6

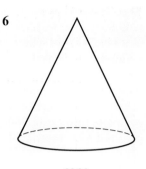

cone

7

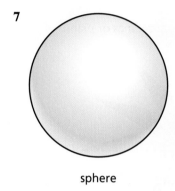

sphere

8

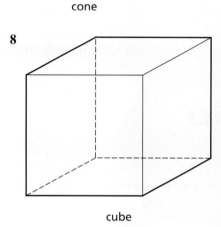

cube

Rotation

If an object is rotated it undergoes a 'turning' movement about a specific point, known as the centre of rotation. An object or shape has rotational symmetry if, during a complete rotation of 360°, it looks the same as it did at the start, at least twice. The number of times a shape looks the same as it did at the start, during a complete turn, is known as the **order of rotational symmetry**.

Example

a What is the order of rotational symmetry of the shape below?
b What is the angle between each of the repeating objects?

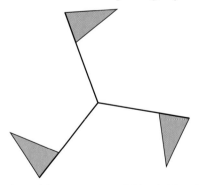

a In one complete rotation, this pattern would look identical three times.
Therefore it has rotational symmetry order 3.
b For the shape to have rotational symmetry, the angles must be equal.
Therefore the angle = 360° ÷ 3 = 120°.

Note. Every shape has rotational symmetry of at least order 1. However, shapes with rotational symmetry of order 1 are not considered to have rotational symmetry. For example, this shape would look identical once in a complete turn (after 360°), but it does not have rotational symmetry.

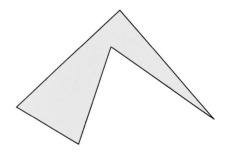

Exercise 14.6

Work out the order of rotational symmetry of each of the following shapes.

1

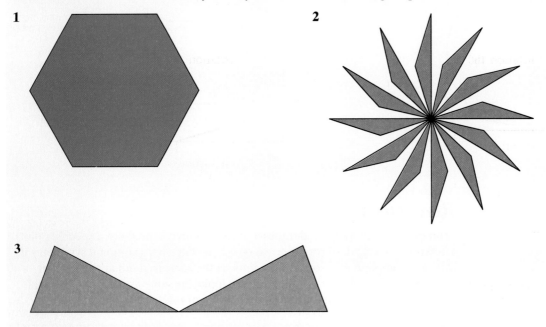

2

3

When describing a rotation, it is necessary to identify not only the position of the centre of rotation, but also the angle of the turn. By convention, unless otherwise stated, the angle is measured **anti-clockwise**.

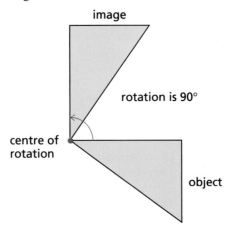

image

rotation is 90°

centre of
rotation

object

Exercise 14.7

In the following questions, the object and the centre of rotation have both been given. Copy each diagram and rotate the object by the amount shown.

1

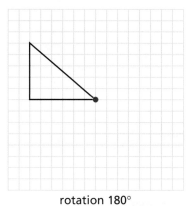

rotation 180°

2

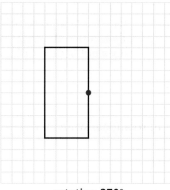

rotation 270°

3

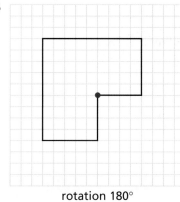

rotation 180°

4

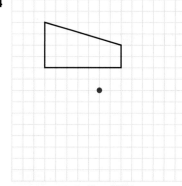

rotation 270°

5

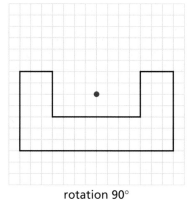

rotation 90°

6

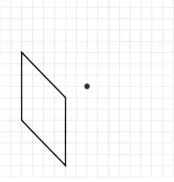

rotation 270°

Measuring the angle of rotation

To measure the angle of rotation, take corresponding points on the object and image and join them to the centre of rotation by straight lines. Measure the angle between the two straight lines; this is the angle of rotation.

Examples   The triangle ABC below is rotated about a point O, to a new position $A'B'C'$. Determine the angle of rotation.

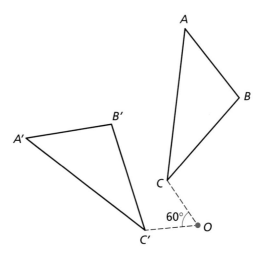

> The angle between OB and OB' would also be 60°.

As can be seen, C on the object has moved to C' on the image. The angle is measured anti-clockwise between OC and OC'.
The angle of rotation is therefore 60°.

Measure the angle of rotation between the object $ABCD$ and the image $A'B'C'D'$ in the diagram below.

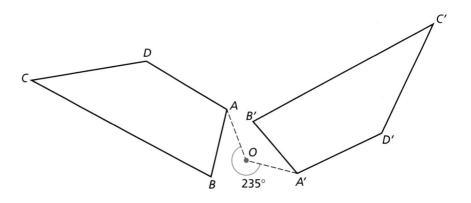

Because rotations are measured in an anti-clockwise direction, the angle between OA and OA' is the reflex angle.
Therefore the angle of rotation is 235°.

Exercise 14.8

You will need:
• protractor or angle measurer

For each of the following diagrams, trace the object, the image and the centre of rotation. Determine the angle of rotation in each case.

1

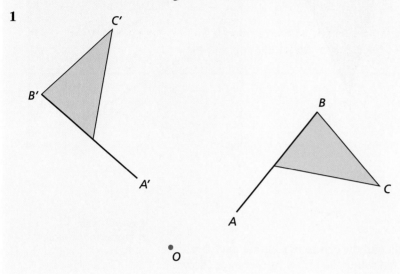

2

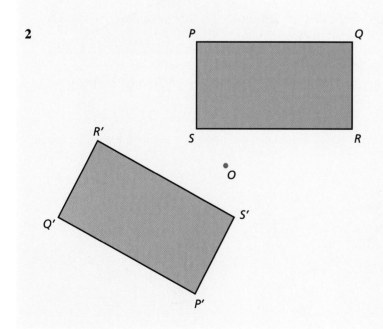

3

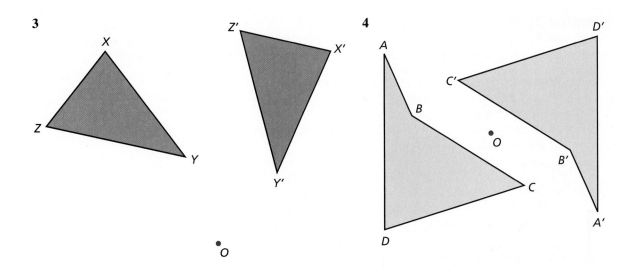

4

Exercise 14.9

In the following questions, the object and the centre of rotation have been drawn. Copy each of the diagrams and rotate the object by the angle shown.

You will need:
• protractor or angle measurer

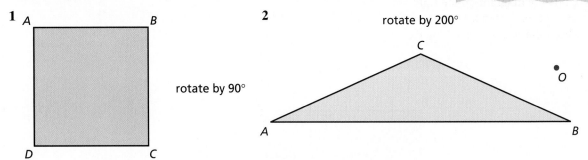

1

rotate by 90°

2

rotate by 200°

3

rotate by 270°

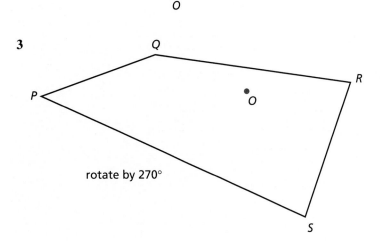

SUMMARY

By the end of this chapter you should know:

■ that **reflections** and **rotations** are types of transformation
■ how to identify the types of symmetry in an object or pattern
■ how to reflect an object about a given **mirror line**
■ that three-dimensional objects can also have **reflective symmetry**
■ how to rotate an object about a point
■ how to measure an angle of rotation.

Exercise 14A

1 Copy the diagram on the right and reflect the triangle in the mirror line shown.

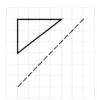

2 Copy the diagram on the right and rotate the trapezium 90° about the centre of rotation *O*.

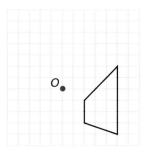

3 Copy the diagram on the right.
  a Draw in the mirror line(s).
  b Find the equation(s) of the mirror line(s).

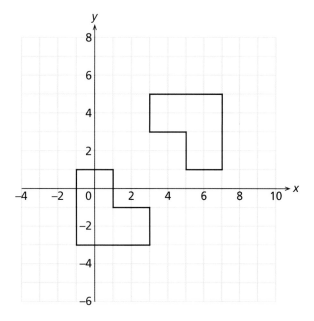

4 In the diagram below, the object ABC is rotated to position $A'B'C'$.
By measurement, find the angle of rotation.

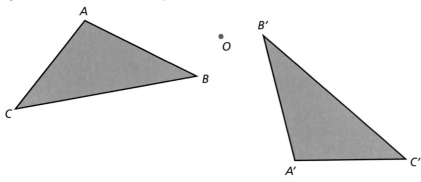

You will need:
- protractor or angle measurer

5 Draw three cubes. On each one draw a different plane of symmetry.

Exercise 14B

1 Copy the diagram on the right and reflect the parallelogram in the mirror line shown.

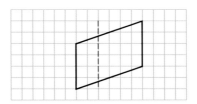

2 Copy the diagram on the right and rotate the quadrilateral 180° about the centre of rotation O.

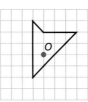

3 Copy the diagram on the right.
  a Draw in the position of a mirror line with equation $y = x - 1$.
  b Reflect the object in the mirror line.

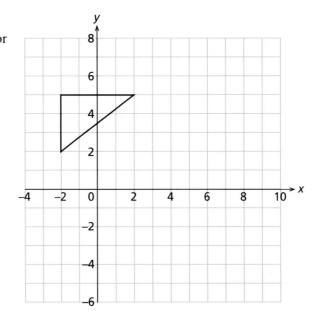

4 Copy the diagram below. Draw the image when the object is rotated by the angle shown.

rotate by 90°

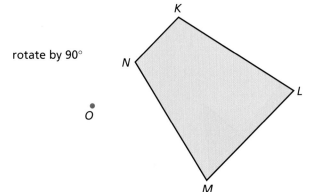

5 Draw three identical cuboids (not cubes). On each one draw a different plane of symmetry.

Exercise 14C

1 For the equilateral triangle below, write down the number of lines of symmetry it has, and also its order of rotational symmetry.

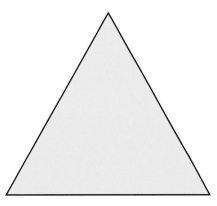

2 For the square below, write down the number of lines of symmetry it has, and also its order of rotational symmetry.

3 Investigate the relationship between the lines of reflective symmetry and the order of rotational symmetry for regular polygons.

4 Does the relationship for regular shapes hold for irregular shapes? Find examples to illustrate your answers.

Exercise 14D

You will need:
• computer with Cabri II
  or similar software
  installed

Using Cabri II, or similar geometry software, produce some patterns
with rotational and/or reflective symmetry.

The example below shows a pattern with rotational symmetry order 8.

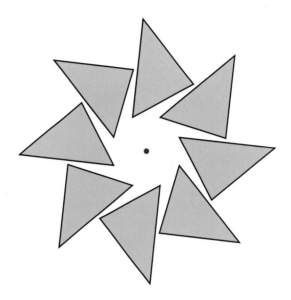

Exercise 14E

Young children often play with a toy called a kaleidoscope. Find out what this toy is, and explain how it
works, with reference to reflective and rotational symmetry.

15 Coordinates

On 22 October 1707 four English war ships, *The Association* (the flagship of Admiral Sir Clowdisley Shovell) and three others, struck the Gilstone Ledges off the Scilly Isles and more than two thousand men drowned. Why? Because the Admiral had no way of knowing exactly where he was. He needed two coordinates to place his position on the sea. He only had one, his latitude.

The story of how to solve the problem of fixing the second coordinate (longitude) is told in Dava Sobel's book *Longitude*. Parliament offered a prize of £20 000 (millions of pounds at today's prices) to anyone who could solve the problem of how to fix longitude at sea.

This chapter is about fixing a position accurately on a line, on a flat surface and in space, that is, in one, two and three dimensions.

Revision

To fix a point in two dimensions (2D), its position is given in relation to a point called the **origin**. Through the origin, axes are drawn perpendicular to each other. The horizontal axis is known as the **x-axis**, and the vertical axis is known as the **y-axis**.

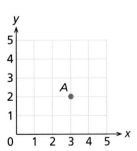

> **Remember:**
> *The idea of describing position using numbers like this was devised by René Descartes, a French philosopher, scientist and mathematician (1596–1650).*

The x-axis is numbered from left to right. The y-axis is numbered from bottom to top.

The position of point A is given by two coordinates: the x-coordinate first, followed by the y-coordinate. So the coordinates of point A are (3, 2).

A number line can extend in both directions by extending the *x*- and *y*-axes below zero, as shown in the grid below.

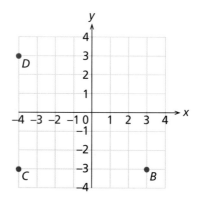

Points *B*, *C* and *D* can be described by their coordinates:

point *B* is at (3, −3)
point *C* is at (−4, −3)
point *D* is at (−4, 3)

Exercise 15.1

You will need:
• squared paper

1 Draw a grid with centre (0, 0), the origin, and mark axes *x* and *y* with scales from −8 to +8. Mark on your grid the points given by the following coordinates.

    a *A* (5, 2)   **b** *B* (7, 3)   **c** *C* (2, 4)
    d *D* (−8, 5)   **e** *E* (−6, −8)   **f** *F* (3, −7)
    g *G* (7, −3)   **h** *H* (6, −6)

For questions 2–4 below:

a draw a separate grid for each question with *x*- and *y*-axes from −6 to +6,
b plot the points, join them in order, and name the shape drawn.

2 *A* (3, 2)   *B* (3, −4)   *C* (−2, −4)   *D* (−2, 2)
3 *E* (1, 3)   *F* (4, −5)   *G* (−2, −5)
4 *H* (−6, 4)   *I* (0, −4)   *J* (4, −2)   *K* (−2, 6)

Exercise 15.2

You will need:
• squared paper

Prepare a coordinate grid with the axes *x* and *y* going from −10 to +10.

1 Plot the points *P* (−6, 4), *Q* (6, 4), *R* (8, −2).
   Plot point *S* such that *PQRS*, when joined, form a parallelogram.
   a Draw diagonals *PR* and *QS*. What are the coordinates of their point of intersection?
   b What is the area of *PQRS*?
2 On the same grid, plot points *M* at (−8, 4) and *N* at (4, 4).
   a Join points *MNRS*. What shape is formed?
   b What is the area of *MNRS*?
   c Explain your answer to **b** above.

3 a On the same grid, plot point *J* where point *J* has *y*-coordinate +10 and *JRS*, when joined, forms an isosceles triangle.
  b What is the *x*-coordinate of all points on the axis of symmetry of triangle *JRS*?

Exercise 15.3

You will need:
• squared paper

1 a On a grid with axes going from −10 to +10 *construct* a regular hexagon *ABCDEF* with centre (0, 0) and where the coordinates of *A* are (0, 8).
  b Write down the approximate coordinates of points *B*, *C*, *D*, *E* and *F*.
2 a On a similar grid, draw an octagon *PQRSTUVW* which has point *P* at (2, −8), point *Q* at (−6, −8) and point *R* at (−10, −4). *PQ* = *RS* = *TU* = *VW* and *QR* = *ST* = *UV* = *WP*.
  b List the coordinates of points *S*, *T*, *U*, *V*, *W*.
  c What are the coordinates of the centre of rotational symmetry of the octagon?

Exercise 15.4

1 The points *A*, *B*, *C* and *D* are not at whole-number points on the number line below. Point *A* is at 0.7.

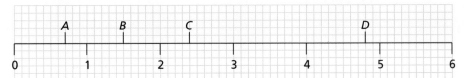

What are the positions of points *B*, *C* and *D*?

2

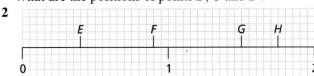

On this number line point *E* is at 0.4 (two small squares represent 0.1). What are the positions of points *F*, *G* and *H*?

3

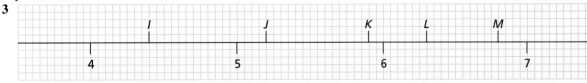

What are the positions of points *I*, *J*, *K*, *L* and *M*? (Each small square is 0.05, i.e. two squares are 0.1.)

4

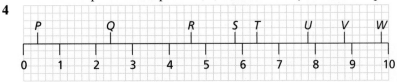

Point *P* is at position 0.4 and point *W* is at position 9.8 (each small square is 0.2).
What are the positions of points *Q*, *R*, *S*, *T*, *U*, *V*?

Exercise 15.5

1 From the coordinate grid below, give the coordinates of points *A*, *B*, *C* and *D*.

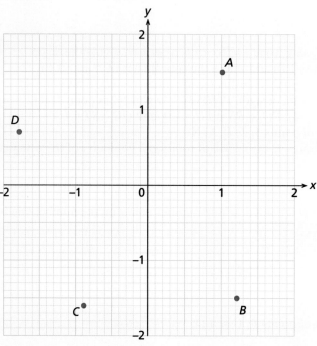

2 From the coordinate grid below, give the coordinates of points *E*, *F*, *G* and *H*.

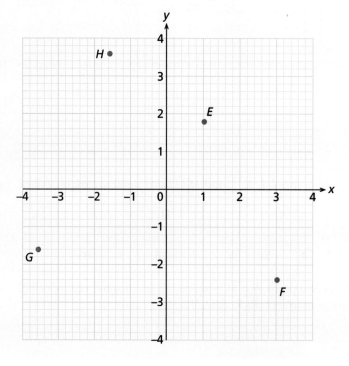

3 From the coordinate grid below, give the coordinates of points *J*, *K*, *L* and *M*.

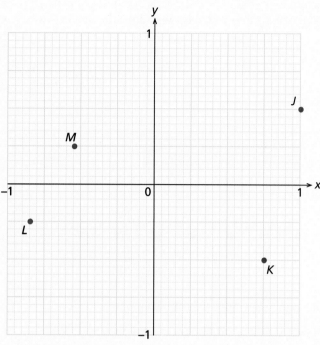

4 From the coordinate grid below, give the coordinates of points *P*, *Q*, *R* and *S*.

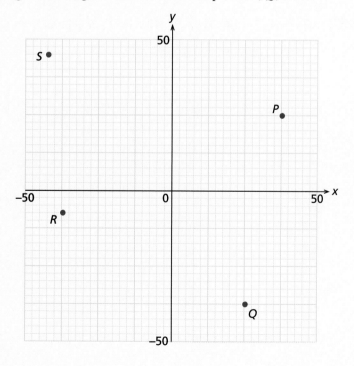

5 Weigh six different British coins in grams and measure their diameter in centimetres (both measurements should be correct to one decimal place). Plot these measurements on a graph with axes similar to the one below.

You will need:
- six different coins
- scales
- ruler
- graph paper

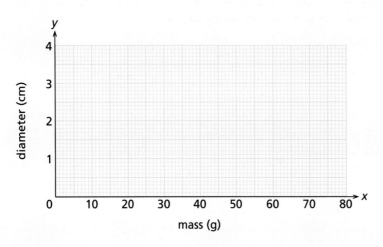

6 On a grid the same as the one below, plot the following points.

A (1.2, 2.8) B (2.5, −1.6) C (−2.8, −0.4) D (−2.4, 1.8)

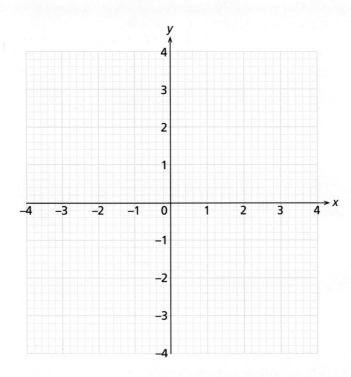

Coordinates in one, two and three dimensions

The position of point A on a number line can be described as its coordinate in one dimension (that dimension being length).

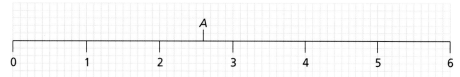

The coordinates of point B are (2, 3); they are the coordinates in two dimensions.

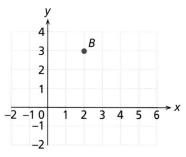

When working in three dimensions, the axes are labelled x, y and z. Each of the axes is at right angles to the other two.

A classroom is in the shape of a cuboid (see the diagram below). Assume that the bottom left-hand corner of the room is the origin. A fly (labelled P) will have coordinates in three dimensions relative to the origin (0, 0, 0).

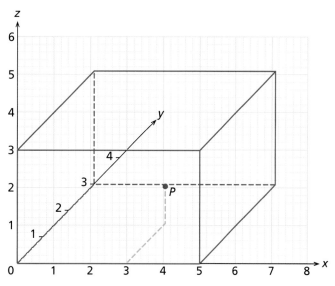

> It is very difficult to illustrate coordinates in three dimensions on the two-dimensional surface of a page.

The coordinates of P are (3, 1.5, 1)

Exercise 15.6

You will need:
• cuboid box

1 Take a shoe box or something similar. Mark axis *x* along its length, axis *y* along its width and axis *z* up its height as shown.

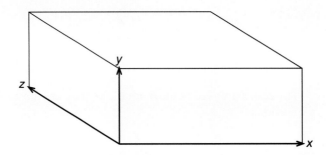

Divide the *x*-axis into ten equal parts and the *y*- and *z*-axes into five equal parts. Indicate to a partner the position of the following points:

A (2, 3, 4) B (7, 4, 1) C (9, 3, 5) D (5, 5, 5) E (8, 3, 0)

2 A cube *ABCDEFGH*, of side length 2 units, is drawn on the axes below.

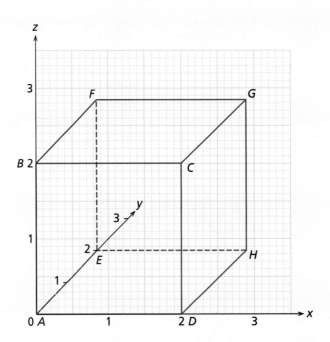

a If corner *A* is positioned at the origin, give the coordinates of each of the corners *A*, *B*, *C*, *D*, *E*, *F*, *G*, *H*.

b What are the coordinates of the centre of the cube?

3 A cuboid *PQRSTUVW* is drawn on the axes below.

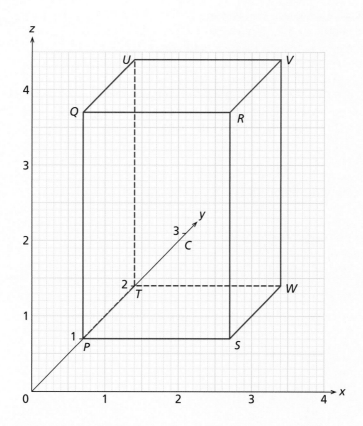

a If corner *P* has coordinates (0, 1, 0), give the coordinates of each of the corners *Q, R, S, T, U, V, W*.
b Describe the position of the point $(2, 1\frac{1}{2}, 1\frac{1}{2})$.

SUMMARY

By the end of this chapter you should know:

- that the position of a point on a grid with *x* (horizontal) and *y* (vertical) axes can be given by two coordinates
- that coordinates can be positive or negative
- that coordinates can be determined to one decimal place by carefully looking at the scale
- that in three dimensions there are three coordinates *x*, *y* and *z* where the three axes *x*, *y* and *z* are at right angles to each other.

Exercise 15A

You will need:
- squared paper
- graph paper

1 Plot these points on an appropriate number line.
   A −6.3 B −5.5 C −0.4 D 2.8 E 5.6
2 Plot these points on a grid with appropriate x- and y-axes.
   F (7, 3) G (5, −3), H (−4, −6) I (−5, 4)
3 Plot these points on a grid with appropriate x- and y-axes.
   P (1.8, 1.4) Q (1.5, −0.8) R (−0.7, −1.2) S (−0.3, 1.4)
4 Show these points on a grid with appropriate x- and y-axes.
   T (40, 35) U (−30, −20) V (−25, 15) W (10, −40)

Exercise 15B

What are the coordinates of the points shown in questions 1–3?

1

2

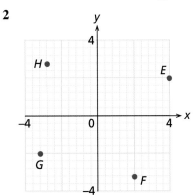

3

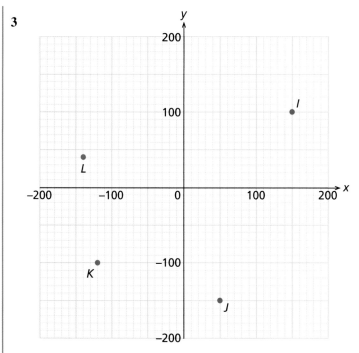

4 The triangular prism *ABCDEF* below is plotted in three dimensions relative to *x*-, *y*- and *z*-axes.

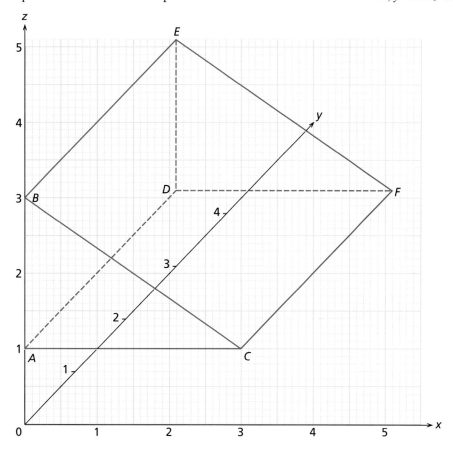

If the coordinates of corner *A* are $(0, 0, 1)$ deduce the coordinates of each of the corners *B*, *C*, *D*, *E* and *F*.

Exercise 15C

- Using the axes on the right, give the coordinates of the corners of triangle 1.
- Write down the coordinates of the corners of triangle 2.
- Write down the coordinates of the corners of triangle 3.
- Without drawing triangle 4, can you predict the coordinates of its corners?
- Can you predict the coordinates of triangle 10?
- What are the coordinates of triangle n?

Show your working clearly.

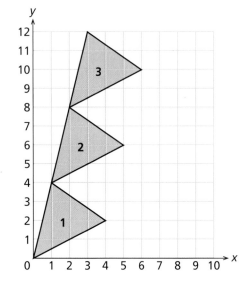

Exercise 15D

You will need:
- graphical calculator

Using a graphical calculator, use the 'draw menu' to draw a coordinate picture of your choice. An example using only positive coordinates is given below.

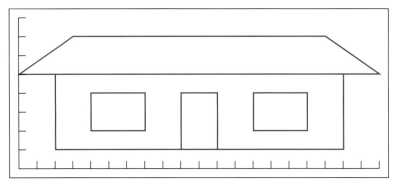

Exercise 15E

In the eighteenth century, the 'longitude problem' was the biggest scientific dilemma of the time (and had been for centuries). Because of sailors' inability to measure their longitude, many had been lost at sea as soon as they lost sight of the land. The search for a solution occupied scientists for two centuries, and in 1714 Parliament offered £20000 (millions of pounds in today's money) to anyone who could provide a successful method or device for measuring longitude.

Dava Sobel's book *Longitude* (published by Fourth Estate, London) describes how the problem was solved. Try to get a copy – it's well worth reading!

The book was dramatised in a two-part television drama; see it if you can.

16 Collecting and analysing data

Who made the following statements?

73% of all statistics are made up on the spot.

Figures can't lie but liars can figure.

There are lies, damned lies and statistics.

In answering that question I was not lying. I was, however, being economical with the truth.

The word 'statistics' comes from the Latin *status*, meaning 'state'. Originally, statistics related to information which was useful to the country or state.

Mathematics has traditionally been seen as searching for truth. Mistakes would be made, but the intention of the argument was to seek clarification and advance knowledge. Statistics is a branch of mathematics that has been brought into disrepute, not by mathematicians, but by people misusing statistical data and methods for their own ends.

The following two statements are deductions made from data available as a result of investing £1000 in 'Pullies', a company that markets knitwear. Both statements are true but each one slants the information it gives for its own purpose.

year 1
£30

year 2
£62

year 3
£125

Statement A:
Profits are booming at 'Pullies'. Profits double each year.

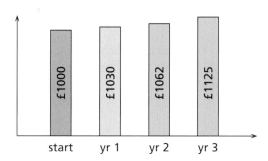

start　yr 1　yr 2　yr 3

£1000　£1030　£1062　£1125

Statement B:
'Pullies' disappoints. Average returns of about 4% could be bettered in a savings account.

Remember:
Diagrams must:

- *have regular scales*
- *start from zero*
- *have equal-width bars for the same interval*
- *have two-dimensional measures (if 3D shapes are used, they should only 'grow' in one direction).*

Discuss:

- how the illustrations are used to justify the statements
- which of the statements is factually true
- possible motives behind the slant on the facts.

Discrete data

Discrete data is data that must take specific values. The number of pupils in a class is an example of discrete data, as the results must be whole numbers (you cannot have 0.3 of a pupil!).

Data can be represented in a **tally** and/or **frequency table**.

Example   The shoe sizes of 10 pupils in a class are written below:

5   7 $6\frac{1}{2}$ 6 $6\frac{1}{2}$ 5   6 $4\frac{1}{2}$ $6\frac{1}{2}$ 7

Enter these results in a tally and frequency table

shoe size	tally	frequency
$4\frac{1}{2}$	/	1
5	//	2
$5\frac{1}{2}$		0
6	//	2
$6\frac{1}{2}$	///	3
7	//	2

Exercise 16.1

1 The figures in the list below give the total number of chocolate buttons in each of 20 packets of buttons.

35   36   38   37   35   36   38   36   37   35
36   36   38   36   35   38   37   38   36   38

Present the figures in a tally and frequency table.

2 A survey is carried out among a group of students to find their favourite subject at school. The results are listed below:

art   maths   science   English   maths   art   English   maths   English   art   science   science   science   maths   art   English   art   science   maths   English   art

Present the results of this survey in a tally and frequency table.

3 a Record the shoe sizes of everybody in your class. Present the results in a tally and frequency table.
b What conclusions can you make from your results?

If there is a big range in the data, it is sometimes easier and more useful to group the data in a **grouped frequency table**.

Example   The scores for the first round of a golf competition are shown below.

71 75 82 96 83 75 76 82 103 85 79 77 83 85 88
104 76 77 79 83 84 86 88 102 95 96 99 102 75 72

Display the above information in a grouped frequency table.

score	frequency
71–75	5
76–80	6
81–85	8
86–90	3
91–95	1
96–100	3
101–105	4
total	30

Note. The groups are arranged so that no score can appear in two groups.

Exercise 16.2

1 The following data gives the percentage scores obtained by students from two classes, 11X and 11Y, in a GCSE maths exam.

11X
42 73 93 85 68 58 33 70 71 85 90 99 41 70 65
80 73 89 88 93 49 50 57 64 78 79 94 80 50 76 99

11Y
70 65 50 89 96 45 32 64 55 39 45 58 50 82 84
91 92 88 71 52 33 44 45 53 74 91 46 48 59 57 95

a Draw a grouped tally and frequency table for each of the classes.
b Illustrate your results on a graph.
c Comment on any similarities or differences from the results.

2 The numbers of apples collected from 50 trees are recorded below.

35   78 15 65   69 32 12   9 89 110 112 148 98
67   45 25 18   23 56 71   62 46 128   7 133 96
24   38 73 82 142 15 98   6 123 49   85   63 19
111 52 84 63   78 12 55 138 102 53   80

a Choose suitable groups for this data and represent it in a grouped frequency table.
b Illustrate your results on a suitable graph.

Average

'Average' is a word which in general use is taken to mean 'somewhere in the middle'. For example, a woman may describe herself as being of average height. A student may think that he or she is of average ability in maths. However, mathematics is more precise, and uses three main methods to measure average.

- The **mode** is the value occurring most often (there can be more than one mode).
- The **median** is the middle value when the data is arranged in order of size.
- The **mean** is found by adding together all the values of the data and then dividing the total by the number of data values.

Example The numbers below represent the goals scored by a football team in the first 15 matches of the season. Find the mean, median and mode of the number of goals.

1 0 2 4 1 2 1 1 2 5 5 0 1 2 3

$$\text{Mean} = \frac{1+0+2+4+1+2+1+1+2+5+5+0+1+2+3}{15} = 2$$

Note. Although in this case the mean is exactly 2, the mean can be a number that is not in the data set at all, for example 2.3 goals per match.

Arranging all the data in order and then picking out the middle number gives the median:

0 0 1 1 1 1 1 ② 2 2 2 3 4 5 5

The mode is the number that appeared most often. Therefore the mode is 1. Had there been one more score of 2, the modes would have been 1 and 2.

Note. If there is an even number of data values, then there will not be one middle number, but a middle pair. The median is calculated by working out the mean of the middle pair.

Exercise 16.3

In questions 1–4, find the mean, median and mode for each set of data.

1 A hockey team plays 15 matches. Below is a list of the number of goals scored in each match.

1 0 2 4 0 1 1 1 2 5 3 0 1 2 2

2 The total scores when two dice are thrown are:

7 4 5 7 3 2 8 6 8 7 6 5 11 9 7 3 8 7 6 5

3 The numbers of pupils present in a class over a three-week period are:

28 24 25 28 23 28 27 26 27 25 28 28 28 26 25

4 An athlete keeps a record (in seconds) of her training times for the 100 m race:

14.0 14.3 14.1 14.3 14.2 14.0 13.9 13.8
13.9 13.8 13.8 13.7 13.8 13.8 13.8

5 The mean mass of the 11 players in a football team is 80.3 kg. The mean mass of the team plus a substitute is 81.2 kg. Calculate the mass of the substitute.

6 After eight matches a basketball player had scored a mean of 27 points. After three more matches his mean was 29. Calculate the total number of points he scored in the last three games.

Large amounts of data

When there are only three values in the sets of data, the median value is given by the second value.

1 ② 3

When there are four values in the set of data, the median value is given by the mean of the second and third value, i.e.

1 ⟨2 3⟩ 4

When there are five values in the set of data, the median value is given by the third value.

If this pattern is continued, it can be deduced that for n values in the set of data, the median value is given by the $\dfrac{n+1}{2}$ value.

This is useful when finding the median of large sets of data.

Example The shoe sizes of 49 people are recorded in the table below. Calculate the median, mean and modal shoe size.

shoe size	3	$3\frac{1}{2}$	4	$4\frac{1}{2}$	5	$5\frac{1}{2}$	6	$6\frac{1}{2}$	7
frequency	2	4	5	9	8	6	6	5	4

As there are 49 values in the set of data, the median value is the 25th value. This occurs within shoe size 5. So the median shoe size is 5.

The mean shoe size is

$$\frac{(3\times 2) + (3\frac{1}{2}\times 4) + (4\times 5) + (4\frac{1}{2}\times 9) + (5\times 8) + (5\frac{1}{2}\times 6) + (6\times 6) + (6\frac{1}{2}\times 5) + (7\times 4)}{49}$$

So the mean shoe size is 5.1 (correct to one decimal place).

The modal shoe size is $4\frac{1}{2}$.

....
Note that the mean value is not necessarily a real shoe size. It does not have to be a member of the data set.
....

Exercise 16.4

1 An ordinary dice was rolled 60 times. The results are shown in the table below. Calculate the mean, median and mode of the scores.

score	1	2	3	4	5	6
frequency	12	11	8	12	7	10

2 Two dice were thrown 100 times. Each time their combined score was recorded. Below is a table of the results. Calculate the mean, median and mode of the scores.

score	2	3	4	5	6	7	8	9	10	11	12
frequency	5	6	7	9	14	16	13	11	9	7	3

3 Sixty flowering bushes are planted. At their flowering peak, the number of flowers per bush is counted and recorded. The results are shown in the table below.

flowers per bush	0	1	2	3	4	5	6	7	8
frequency	0	0	0	6	4	6	10	16	18

a Calculate the mean, median and mode of the number of flowers per bush.
b Which of the mean, median and mode would be most useful when advertising the bush to potential buyers?

Mean and mode for grouped data

As has already been described, sometimes it is more useful to group data, particularly if the range of values is very large. However, by grouping data some accuracy is lost. The results below are the distances (to the nearest metre) run by 20 pupils in one minute.

256 271 271 274 275 276 276 277 279 280
281 282 284 286 287 288 296 300 303 308

Table 1: Class interval of 5 m

group	250–254	255–259	260–264	265–269	270–274	275–279	280–284	285–289	290–294	295–299	300–304	305–309
frequency	0	1	0	0	3	5	4	3	0	1	2	1

Table 2: Class interval of 10 m

group	250–259	260–269	270–279	280–289	290–299	300–309
frequency	1	0	8	7	1	3

Table 3: Class interval of 20 m

group	250–269	270–289	290–309
frequency	1	15	4

The three tables above highlight the effects of different group sizes. Table 1 is perhaps too detailed, while in Table 3 the group sizes are too big and so most of the results fall into one group. Table 2 is the most useful, in that the spread of the results is still clear; however, detail is lost. In the 270–279 group we can see that there are 8 pupils, but without the raw data we would not know where in the group they lie. To find the mean of **grouped data** we assume that all the data within a group takes the **mid-interval value**, for example

group	250–259	260–269	270–279	280–289	290–299	300–309
mid-interval value	254.5	264.5	274.5	284.5	294.5	304.5
frequency	1	0	8	7	1	3

$$\text{mean} = \frac{(254.5 \times 1) + (264.5 \times 0) + (274.5 \times 8) + (284.5 \times 7) + (294.5 \times 1) + (304.5 \times 3)}{20}$$

$$= \frac{5650}{20}$$

$$= 282.5$$

The **estimate** of the mean distance run is 282.5 m.
The modal group is 270–279.
Note that the mode is called the **modal group** for grouped data.

Exercise 16.5

1 A pet shop has 100 tanks containing fish. The number of fish in each tank is recorded in the table below.

number of fish	0–9	10–19	20–29	30–39	40–49
frequency	7	12	24	42	15

  a Calculate an estimate for the mean number of fish in each tank.
  b Give the modal group.

2 A school has 148 year 11 pupils. Their percentage scores in their mathematics mock exam are recorded in the grouped frequency table below.

percentage score	0–9	10–19	20–29	30–39	40–49	50–59	60–69	70–79	80–89	90–99
frequency	3	2	4	6	8	36	47	28	10	4

  a Calculate an estimate for the mean percentage score for the mock exam.
  b What was the modal group score?

3 A station master records how many minutes late each train is to the nearest minute. The table of results is given below.

number of minutes late	0–4	5–9	10–14	15–19	20–24	25–29
frequency	16	9	3	1	0	1

  a Calculate an estimate for the mean number of minutes late for the trains.
  b What is the modal number of minutes late?
  c The station master's report concludes 'The average number of minutes late is less than 5.' Comment on this conclusion.

Data collection

There are several ways of collecting data on a particular topic. One of the most commonly used ways is a questionnaire. Questionnaires are a useful tool, as a lot of data can be collected very quickly. There are two main types of questionnaire; sometimes the interviewer asks someone the questions and makes a note of the replies, at other times people fill in the questionnaire themselves. However, to carry out a successful survey, the design of the questionnaire needs to be given some thought.

- Always have a clear aim for your survey, i.e. what exactly are you trying to find out?
- Questions should not be **biased**.
- Questions should always give several possible answers to choose from.
- Do not ask questions that are likely to upset or embarrass people.
- Stick to questions that are relevant.
- Ask questions that are clear.
- Try to have the questions in a logical order.

Exercise 16.6

A pupil is conducting a survey to do with health. Use the questionnaire guidelines to explain what is wrong with each of the following questions.

1 Are you fat?
2 Sweets are really bad for your teeth. Do you eat lots of sweets?
3 Do you eat apples, bananas and grapes?
4 Where do you do most of your food shopping?
  Tesco ☐   Sainsbury ☐
5 What is your favourite colour?
6 If healthy food was cheaper, would you eat more healthily?
7 Do you have bad breath?

Exercise 16.7

You will need:
- newspapers

1 Using the questionnaire guidelines given earlier, carry out a survey of your choice. Write a brief report on the findings of your survey.
2 Using newspapers as a resource, find articles that describe the findings of surveys. Draw up a questionnaire that could have been used to collect the data.

SUMMARY

By the time you have completed this chapter you should know:

- what **discrete data** is
- how to draw up a **grouped frequency table**
- how to calculate the three types of average – **mean**, **median** and **mode**
- that calculating the mean of grouped data only gives an approximation
- how to design a questionnaire and carry out a survey.

Exercise 16A

1 Find the mean, median and mode of each of the following sets of data.
   a 63   72   72   84   86
   b 6   6   6   12   18   24
   c 5   10   5   15   5   20   5   25   15   10
2 The mean mass of the 15 players in a rugby team is 85 kg. The mean mass of the team plus a substitute is 83.5 kg. Calculate the mass of the substitute.
3 Thirty families live in a street. The number of children in each family is given in the table below.

number of children	0	1	2	3	4	5	6
frequency	3	5	8	9	3	0	2

   a Calculate the mean number of children per family.
   b Calculate the median number of children per family.
   c Calculate the modal number of children.
4 The numbers of people attending 30 screenings of a film at a local cinema are given below.

   21   30   66   71   10   37   24   21   62   50   27   31   65   12   38
   34   53   34   19   43   70   34   27   28   52   57   45   25   30   39

   a Using groups 10–19, 20–29, 30–39, etc. present the above data in a grouped frequency table.
   b Using your grouped data, calculate an estimate for the mean number of people attending each screening.
5 A supermarket is hoping to carry out a survey to find out what improvements customers would like to see introduced in the store. Write down how the first five questions of the questionnaire might look.

Exercise 16B

1 Find the mean, median and mode of each of the following sets of data.
   a 4   5   5   6   7
   b 3   8   12   18   18   24
   c 4   9   3   8   7   11   3   5   3   8
2 The mean mass of the 11 players in a football team is 76 kg. The mean mass of the team plus a substitute is 76.2 kg. Calculate the mass of the substitute.
3 Thirty children were asked about the number of pets they had. The results are shown in the table below.

number of pets	0	1	2	3	4	5	6
frequency	5	5	3	7	3	1	6

   a Calculate the mean number of pets per child.
   b Calculate the median number of pets per child.
   c Calculate the modal number of pets.

4 The numbers of people attending a disco at a club over 30 evenings are:

89   94   32   45   57   68   127   138   23   77   99   47   44   100   106
132   28   56   59   49   96   103   90   84   136   38   72   47   58   110

  a Using groups 0–19, 20–39, 40–59, etc. present the above data in a grouped frequency table.
  b Using your grouped data, calculate an estimate for the mean number of people going to the disco each night.

5 An estate agent is hoping to carry out a survey by telephone, to find out about the availability of property in an area. Write down five suitable questions for the questionnaire.

Exercise 16C

A recent magazine article stated 'Tabloid newspapers are easier to read than broadsheet newspapers.' Carry out a survey to see if this statement is true.

You will need:
● tabloid and broadsheet newspapers

Exercise 16D

You will need:
● computer with spreadsheet package installed

The Millennium Dome in London was open to the public from 1 January 2000. Below is a list of the number of people who visited the Dome each day during January 2000.

Saturday	1 Jan.	11190	Sunday	2 Jan.	22252
Monday	3 Jan.	21718	Tuesday	4 Jan.	17535
Wednesday	5 Jan.	9414	Thursday	6 Jan.	7506
Friday	7 Jan.	9838	Saturday	8 Jan.	20541
Sunday	9 Jan.	16325	Monday	10 Jan.	6670
Tuesday	11 Jan.	5990	Wednesday	12 Jan.	6243
Thursday	13 Jan.	5084	Friday	14 Jan.	7650
Saturday	15 Jan.	17619	Sunday	16 Jan.	15090
Monday	17 Jan.	7406	Tuesday	18 Jan.	7145
Wednesday	19 Jan.	6918	Thursday	20 Jan.	6794
Friday	21 Jan.	9079	Saturday	22 Jan.	18040
Sunday	23 Jan.	16262	Monday	24 Jan.	6746
Tuesday	25 Jan.	8437	Wednesday	26 Jan.	7696
Thursday	27 Jan.	8208	Friday	28 Jan.	13247
Saturday	29 Jan.	23304	Sunday	30 Jan.	19486
Monday	31 Jan.	6987			

Use a spreadsheet and appropriate formulae to answer the following questions.

1 What was the total number of visitors to the Dome during January?
2 What was the mean daily number of visitors?
3 What was the mean number of visitors for each of the different days of the week?
4 Give an explanation for the differences in mean attendance in question 3.
5 Suggest reasons why this information may be useful to managers at the Dome.

One possible way of setting out your spreadsheet is given below:

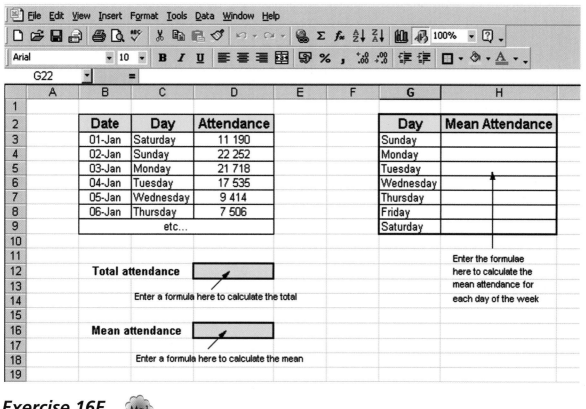

Exercise 16E

What is a 'Gallup Poll', and why is it so called? When and why was the first Gallup Poll conducted?

17 Representing data

In chapter 16 we looked at how to analyse data in a variety of ways. Being able to analyse data is important, as raw data by itself is not usually presented in a way that makes it easy for the reader to draw conclusions. One way to make data easier to understand is to display it graphically.

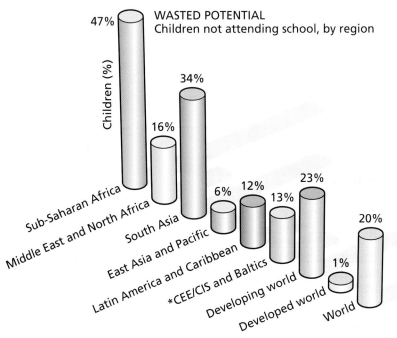

WASTED POTENTIAL
Children not attending school, by region

*Central & Eastern Europe/Commonwealth of Independent States
Source: adapted from UNICEF statistics, 1997

OIL

The world currently relies on fossil fuels –
oil, coal and gas – for most of its energy

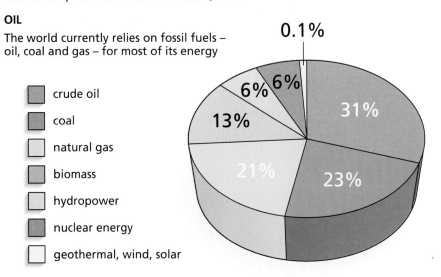

- crude oil
- coal
- natural gas
- biomass
- hydropower
- nuclear energy
- geothermal, wind, solar

In this chapter we will look at some of the ways in which data can be displayed graphically.

Frequency diagrams

Frequency diagrams are one of the more common types of graph. They involve the use of bars for each category. The height of each bar represents the frequency of each category. For example, the table below shows the number of brothers and sisters of 30 pupils in a class.

number of brothers and sisters	0	1	2	3	4	5
frequency	5	8	11	3	2	1

This can be represented on a frequency diagram as follows:

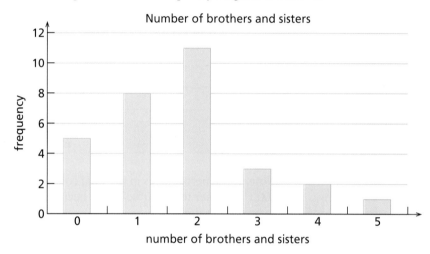

Number of brothers and sisters

Pie charts

Pie charts are another common way of representing data. The data above can be displayed on a pie chart as shown below.

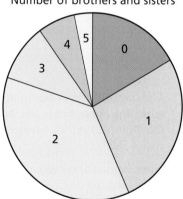

Number of brothers and sisters

In a pie chart, the size of each **sector** (slice) is proportional to the frequency of each category. The size of each sector is calculated either by using percentages or by using degrees.

Example   Look again at the data for the number of brothers and sisters of pupils in a class.

number of brothers and sisters	0	1	2	3	4	5
frequency	5	8	11	3	2	1

a Express the frequency of each category as a percentage.
b Express the frequency of each category in degrees.

a Total number of pupils is $5 + 8 + 11 + 3 + 2 + 1 = 30$.

number of brothers and sisters	number of brothers and sisters as a fraction of total	percentage
0	$\frac{5}{30}$	$\frac{5}{30} \times 100 = 16.7\%$
1	$\frac{8}{30}$	$\frac{8}{30} \times 100 = 26.7\%$
2	$\frac{11}{30}$	$\frac{11}{30} \times 100 = 36.7\%$
3	$\frac{3}{30}$	$\frac{3}{30} \times 100 = 10\%$
4	$\frac{2}{30}$	$\frac{2}{30} \times 100 = 6.7\%$
5	$\frac{1}{30}$	$\frac{1}{30} \times 100 = 3.3\%$

Note. If the percentages in the final column are totalled, they come to 100.1%, rather than 100%. This is because each percentage has been rounded to one decimal place.

b Total number of pupils is $5 + 8 + 11 + 3 + 2 + 1 = 30$.

number of brothers and sisters	number of brothers and sisters as a fraction of total	degrees
0	$\frac{5}{30}$	$\frac{5}{30} \times 360 = 60°$
1	$\frac{8}{30}$	$\frac{8}{30} \times 360 = 96°$
2	$\frac{11}{30}$	$\frac{11}{30} \times 360 = 132°$
3	$\frac{3}{30}$	$\frac{3}{30} \times 360 = 36°$
4	$\frac{2}{30}$	$\frac{2}{30} \times 360 = 24°$
5	$\frac{1}{30}$	$\frac{1}{30} \times 360 = 12°$

The only difference between **a** and **b** in the example is that, as **a** is a percentage table, the multiplier is 100, because we want the amount to be out of 100. In **b**, the multiplier is 360, because a full circle is 360° and we want the amount to be out of 360°.

Note. A pie chart can be drawn using percentages, with a percentage pie-chart scale, or using degrees, with an angle measurer. It is important to realise that whichever method is used, the pie charts produced are *exactly* the same.

Exercise 17.1

1 A milkman delivers milk to 150 houses on his round. The number of bottles he delivers to each house is shown in the table below.

number of bottles	1	2	3	4	5	6
frequency	27	54	34	16	15	4

Represent the above information in a frequency chart.

2 80 pupils sat a maths exam. On the way out of the exam hall, they were asked whether they had found the exam 'easy', 'OK' or 'hard'. Their responses are shown in the table below.

response	easy	OK	hard	no comment
frequency	16	36	8	20

Represent the above data in a percentage pie chart. Show your workings clearly.

3 During a television game show called 'Who wants to win a lot of money?', a contestant asks the audience to answer a particular question. There are four options: A, B, C and D. 42 people suggest answer A, 535 suggest B, 123 suggest C and 20 suggest D.
   a Assuming all of the audience gave an answer, calculate the total number of people in the audience.
   b What percentage of the audience suggested each of the answers A, B, C and D?
   c Calculate the angle represented by each of the answers A, B, C and D.
   d Represent the audience figures in a pie chart.

4 An ice cream vendor keeps a record of the number and types of ice creams she sells during a weekend in April. The results are shown in the table below.

ice cream	Cornutto	Soliro	Clippo	Megnum	Maximilk	Fib
frequency	52	31	60	18	7	12

   a Calculate the total number of ice creams sold.
   b In this case, for which type of pie chart would it be easier to calculate the size of the sectors? Give a reason for your answer.
   c Represent the above information in a pie chart.

One of the ice cream manufacturers has an advertising campaign just before the summer. The table below shows the number and type of ice creams sold during a weekend in August.

ice cream	Cornutto	Soliro	Clippo	Megnum	Maximilk	Fib
frequency	130	75	163	96	15	21

   d Calculate the total number of ice creams sold.
   e In this case, for which type of pie chart would it be easier to calculate the size of the sectors? Give a reason for your answer.
   f Represent the above information in a pie chart.
   g Which ice cream is most likely to have been advertised? Give a reason for your answer.

5 A fast food restaurant sells four different flavours of milkshake: vanilla, strawberry, chocolate and banana. In one lunchtime, 55% of the milkshakes sold are vanilla flavoured, 25% are strawberry, 15% are chocolate and 5% banana.
   a Convert each of the above percentages into degrees out of 360°.
   b Draw a pie chart of the types of milkshake sold.
   c If a total of 420 milkshakes were sold, calculate how many of each flavour were sold.

You will need:
- squared/graph paper
- compasses
- protractor
- ruler

Remember:
Frequency charts must:
- *have regular scales*
- *start from zero*
- *have equal-width bars for the same interval.*

Scatter graphs

Scatter graphs are particularly useful if we wish to see if there is a **correlation** (relationship) between two sets of data. The two values of data collected represent the coordinates of each point plotted. How the points lie when plotted indicates the type of relationship between the two sets of data.

Example   The heights and weights (masses) of 20 children under the age of five were recorded. The heights were recorded in centimetres and the weights in kilograms. The data is shown below.

height	32	34	45	46	52
weight	5.834	3.792	9.037	4.225	10.149
height	59	63	64	71	73
weight	6.188	9.891	16.010	15.806	9.929
height	86	87	95	96	96
weight	11.132	16.443	20.895	16.181	14.000
height	101	108	109	117	121
weight	19.459	15.928	12.047	19.423	14.331

a Plot a scatter graph of the above data.
b Comment on any relationship you see.
c If another child was measured as having a height of 80 cm, approximately what weight would you expect him or her to be?

a

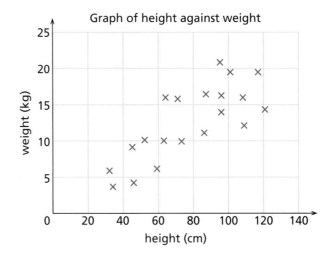

> A line of best fit need not pass through the origin.

> **Remember:**
> *Values predicted from a line of best fit are only approximate – if someone else's line is slightly different from yours, their predictions will also be slightly different.*

b The points tend to lie in a diagonal direction from bottom left to top right. This suggests that as height increases then, in general, weight increases too. Therefore there is a **positive correlation** between height and weight.

c We assume that this child will follow the trend set by the other 20 children. To deduce an approximate value for the weight, we draw a **line of best fit**. This is a solid straight line which passes through the points as closely as possible, as shown below.

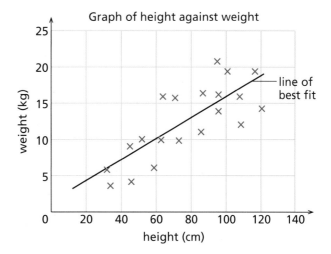

The line of best fit can now be used to give an approximate solution to the question. If a child has a height of 80 cm, you would expect his or her weight to be in the region of 13 kg.

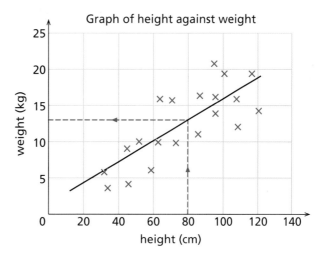

Types of correlation

There are several types of correlation, depending on the arrangement of the points plotted on the scatter graph. These are described below.

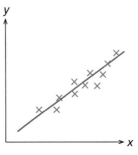

A **strong positive correlation** between the variables x and y. The points lie very close to the line of best fit. As x increases, so does y.

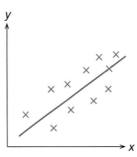

A **weak positive correlation**. Although there is direction to the way the points are lying, they are not tightly packed around the line of best fit. As x increases, y tends to increase too.

No correlation. As there is no pattern to the way in which the points are lying, there is no correlation between the variables x and y. As a result there can be no line of best fit.

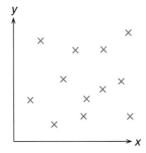

A **strong negative correlation**. The points lie close around the line of best fit. As x increases, y decreases.

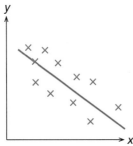

A **weak negative correlation**. The points are not tightly packed around the line of best fit. As x increases, y tends to decrease.

Exercise 17.2

You will need:
- graph paper

or

- computer with spreadsheet package installed

1 State what type of correlation you might expect, if any, if the following data was collected and plotted on a scatter graph. Give reasons for your answer.
  a A student's score in a maths exam and their score in a science exam
  b A student's hair colour and the distance they have to travel to school
  c The outdoor temperature and the number of cold drinks sold by a shop
  d The age of a motorcycle and its second-hand selling price
  e The number of people living in a house and the number of rooms the house has
  f The number of goals your opponents score and the number of times you win
  g A child's height and the child's age
  h A car's engine size and its fuel consumption
2 A newspaper gives daily readings for the number of hours of sunshine and the amount of rainfall in millimetres for several places in the UK. The table below is a summary.

place	hours of sunshine	rainfall (mm)
Aberdeen	9.8	0.3
Aviemore	2.8	0
Belfast	4.1	6.6
Birmingham	6.4	0.8
Bournemouth	9.3	4.3
Bristol	10.2	1.8
Cardiff	10.8	2.8
Folkestone	10.5	0
Hastings	7.2	0.3
Isle of Man	9.3	2.5
Isle of Wight	10.7	0.3
London	10.1	1.5
Manchester	4.6	2.0
Margate	10.9	0
Newcastle	8.2	0.5
Newquay	4.1	2.3
Oxford	10.1	3.0
Scarborough	10.6	1.8
Skegness	12.0	3.3
Southport	8.2	7.4
Torquay	8.8	1.5

  a Plot a scatter graph of hours of sunshine against amount of rainfall. Use a spreadsheet if possible.
  b What type of correlation, if any, is there between the two variables? Comment on whether this is what you would expect.

3 The United Nations keep an up-to-date database of statistical information on its member countries. The table below shows some of the information available.

country	life expectancy at birth (years, 1990–99)		adult illiteracy rate (%, 1995)	infant mortality rate (per 1000 births, 1990–99)
	female	male		
Australia	81	76	0	6
Barbados	79	74	2.6	12
Brazil	71	63	16.8	42
Chad	49	46	51.9	112
China	72	68	18.5	41
Colombia	74	67	9.6	30
Congo	51	46	25.6	90
Cuba	78	74	4.4	9
Egypt	68	65	48.9	51
France	82	74	0	6
Germany	80	74	0	5
India	63	62	48	72
Iraq	64	61	42	95
Israel	80	76	4.9	8
Japan	83	77	0	4
Kenya	53	51	22.7	66
Mexico	76	70	10.5	31
Nepal	57	58	64.1	83
Portugal	79	72	10	9
Russian Federation	73	61	0.9	18
Saudi Arabia	73	70	27.8	23
United Kingdom	80	75	0	7
United States of America	80	73	0	7

a By plotting a scatter graph, decide if there is a correlation between the adult illiteracy rate and the infant mortality rate.

b Are your findings in part a what you expected? Explain your answer.

c Without plotting a graph, decide if you think there is likely to be a correlation between male and female life expectancy at birth. Explain your reasons.

d Plot a scatter graph to test if your predictions for part c were correct.

SUMMARY

By the time you have completed this chapter you should know:

- what a **frequency diagram** is and how it is constructed (see page 203)
- how to draw a **pie chart** (see page 203)
- how to plot a **scatter graph** (see page 206)
- how to identify the different types of **correlation** (e.g. **positive** or **negative correlation**)
- how to draw a **line of best fit** and use it to predict values (see page 207).

Exercise 17A

You will need:
- squared/graph paper
- ruler
- protractor
- compasses

1 The table below gives the average time taken for 30 pupils in a class to get to school each morning, and the distance they live from the school.

distance (km)	2	10	18	15	3	4	6	2	25	23	3	5	7	8	2
time (min)	5	17	32	38	8	14	15	7	31	37	5	18	13	15	8
distance (km)	19	15	11	9	2	3	4	3	14	14	4	12	12	7	1
time (min)	27	40	23	30	10	10	8	9	15	23	9	20	27	18	4

a Plot a scatter graph of distance travelled against time taken.
b Describe the correlation between the two variables.
c Explain why some pupils who live further away may get to school more quickly than some of those who live nearer.
d Draw a line of best fit on your scatter graph.
e A new pupil joins the class. Use your line of best fit to estimate how far away from school she might live if she takes, on average, 19 minutes to get to school each morning.

2 The 1997 General Election results in Cambridge were as follows:

Labour	27 436
Conservative	13 299
Liberal Democrat	8 287
Referendum Party	1 262
Green Party	654
Other	401

a How many people voted in Cambridge in the 1997 General Election?
b What percentage of voters voted for Labour?
c Draw a pie chart of the results. Show your working clearly.

Exercise 17B

1 All the pupils in year 10 in a school were asked what their favourite type of music was. The results are given in the table below.

type of music	number of pupils
dance	53
rock	38
pop	27
indie	40
rap	8
classical	10
other	4

    a What is the total number of pupils surveyed?
    b Draw a pie chart of the results. Show your workings clearly.

2 A department store decides to investigate if there is a correlation between the number of pairs of gloves it sells and the outside temperature. Over a one-year period the store records, every two weeks, how many pairs of gloves are sold and the mean daytime temperature during the same period. The results are given in the table below.

mean temperature (°C)	3	6	8	10	10	11	12	14	16	16	17	18	18
number of pairs of gloves	61	52	49	54	52	48	44	40	51	39	31	43	35
mean temperature (°C)	19	19	20	21	22	22	24	25	25	26	26	27	28
number of pairs of gloves	26	17	36	26	46	40	30	25	11	7	3	2	0

    a Plot a scatter graph of mean temperature against number of pairs of gloves.
    b What type of correlation is there between the two variables?
    c How might this information be useful for the department store in the future?

Exercise 17C

Investigate one of the following statements.

1 People with big hands have big feet.
2 Pupils who are good at maths are also good at science.
3 People with tall parents are tall themselves.
4 The more time spent playing computer games, the less homework you do.
5 People who wear glasses or contact lenses are better at maths.

Explain your findings clearly, with the aid of appropriate graphs.

Exercise 17D

Elections in Britain are held approximately every five years. During that time the voting habits of people change.

You will need:
- computer with internet access and a spreadsheet package installed

a Using the internet, find out the results of the voting habits of people in your area during the last two General Elections.
b Enter the results into a spreadsheet.
c Using the graphing facility of the spreadsheet package, draw appropriate graphs to demonstrate the changes in voting habits for people in your area for the last two General Elections.

Exercise 17E

A recent survey showed that large housing estates where the per capita income (income per person) was low also had high crime rates. A newspaper editorial concluded that 'it is obvious from the results that poverty in cities causes crime'. Comment on these conclusions with reference to the work you have done on surveys and on correlation.

Per capita means per head or per person.

18 Probability

Revision

Tom Stoppard's play *Rosencrantz and Guildenstern are Dead* sets its tone from a probability experiment in Act 1.

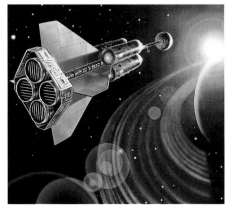

In the book *The Hitchhiker's Guide to the Galaxy*, the spaceship piloted by Zaphod Beeblebrox zipped around the universe with the power of its improbability.

The financier George Soros made £1 billion in a few days by backing the 'balance of probabilities'.

The modern state of Monaco in the South of France was founded on the fact that a little rolling ball is as likely to stop in one slot as in another.

You will remember that probability is the study of chance. **Theoretical probability** is a way of predicting what *should* happen. It is no guarantee of what *will* happen. Take, for example, the outcome of a single roll of a dice.

A **favourable outcome** refers to an event actually occurring.

The total number of **possible outcomes** refers to all the different outcomes one can get in a particular situation.

For example, if you need to throw a 3 with an ordinary dice, there is one favourable outcome (a 3) and the total number of possible outcomes is six (i.e. 1, 2, 3, 4, 5 or 6).

Remember:
If there are n equally likely events, the probability of each one happening is $\frac{1}{n}$.

$$\text{The probability of an event} = \frac{\text{the number of favourable outcomes}}{\text{the total number of equally likely outcomes}}$$

Albert Einstein hated the idea that the universe might be based on random chance – he once said at a conference, 'Gentlemen, God does not play dice.'

Therefore, in the example above, the probability of getting a 3 is P(3) = $\frac{1}{6}$.

You will remember that an estimate of the likelihood of an event happening can be shown on a **probability scale** as shown below.

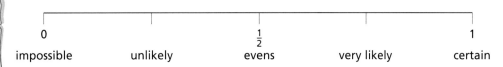

Zero on the line represents an event that is impossible; 1 on the line represents an event that is certain.

Example   An ordinary, fair dice is rolled. Calculate the theoretical probability of getting an even number.

Number of favourable outcomes = 3 (i.e. getting either 2, 4 or 6).
Total number of possible outcomes = 6 (i.e. 1, 2, 3, 4, 5 and 6).
Probability of getting an even number = $\frac{3}{6} = \frac{1}{2}$.

Exercise 18.1

You will need:
• ruler

1 Draw a probability line 12 cm long. At the extreme left put zero and at the extreme right put 1. Write each of the following common words/phrases on the line in the place which best matches their meaning.
   a certain   **b** 50/50   **c** poor chance
   d likely   **e** very likely **f** uncertain
   g no chance

2 Draw a probability line 12 cm long. At the extreme left put zero and at the extreme right put 1. Try to estimate the chances of the following events happening and place them on your line.
   a You will watch TV tonight.
   b You will play sport tomorrow.
   c You will miss school one day in the next month.
   d You will fly on a plane next year.
   e You will learn another language one day.
   f You will do some homework tonight.

3 Calculate the theoretical probability, when rolling an ordinary, fair dice, of getting each of the following.
   a a score of 1   **b** a score of 5   **c** an odd number
   d a score less than 6   **e** a score of 7   **f** a score less than 7
   g a prime number

4 250 tickets are sold for a raffle. What is the probability of winning if you buy:
   a 1 ticket   **b** 5 tickets   **c** 250 tickets
   d 0 tickets

5 In a class there are 25 girls and 15 boys. The teacher takes in all of their books in a random order.
   a Calculate the probability that the teacher will mark a book belonging to a girl first.
   b Calculate the probability that the teacher will mark a book belonging to a boy first.
   c What is the total of **a** and **b**? Explain your answer.

6 Tiles, each lettered with one different letter of the alphabet, are put into a bag. If one tile is drawn out at random, calculate the probability that it is:
   a an A or P   **b** a vowel   **c** a consonant
   d an X, Y or Z   **e** a letter in your first name

7 A boy was late for school 5 times in the last 30 school days. If tomorrow is a school day, calculate the probability that he will arrive late. (Assume there is no pattern in the days on which he arrived late.)

8 **a** Three red, 10 white, 5 blue and 2 green counters are put into a bag. If one is picked at random, calculate the probability that it is:
 i) a green counter   ii) a blue counter
 b If the first counter taken out is green and it is not put back, calculate the probability that the second counter picked is:
 i) a green counter   ii) a red counter

9 A roulette wheel has the numbers 0 to 36 equally spaced around its edge. Assuming that it is unbiased, calculate the probability of spinning it and getting:
 a the number 5   **b** an even number   **c** an odd number
 d zero   **e** a number greater than 15
 f a multiple of 3   **g** a multiple of 3 or 5   **h** a prime number

10 A normal pack of playing cards contains 52 cards. These are made up of four suits (Hearts, Diamonds, Clubs and Spades). Each suit consists of 13 cards. These are labelled Ace, 2, 3, 4, 5, 6, 7, 8, 9, 10, Jack, Queen and King. The Hearts and Diamonds are red; the Clubs and Spades are black. If a card is picked at random from a normal pack of cards, calculate the probability of picking:
 a a Heart   **b** a black card   **c** a 4
 d a red King   **e** a Jack, Queen or King   **f** the Ace of Spades
 g an even-numbered card   **h** a 7 or a Club

If the theoretical probability is known and either the number of favourable outcomes or the number of possible outcomes is known, then the formula for theoretical probability given earlier can be used to find the unknown value.

Examples   250 tickets are sold for a raffle. A girl calculates that the tickets bought by her family give them a 0.032 probability of winning first prize. How many tickets did the family buy?

$$\text{probability} = \frac{\text{number of favourable outcomes } (F)}{\text{total number of possible outcomes}}$$

So

$$0.032 = \frac{F}{250}$$

$$250 \times 0.032 = F$$

$$8 = F$$

The family bought 8 tickets.

A man has 8 tickets for a raffle. His son knows how many tickets have been sold, and tells his father that he has a probability of 0.016 of winning the first prize. How many tickets have been sold?

$$\text{probability} = \frac{\text{number of favourable outcomes}}{\text{total number of possible outcomes } (T)}$$

$$0.016 = \frac{8}{T}$$

$$T \times 0.016 = 8$$

$$T = \frac{8}{0.016}$$

$$T = 500$$

So 500 tickets have been sold.

Exercise 18.2

1 A boy calculates that he has a probability of 0.08 of winning first prize in a raffle. If 500 tickets are sold, how many has he bought?

2 The probability of winning first prize in a spinner game is given as 0.04 for each number chosen. How many numbers are there on the spinner?

3 A bag contains 7 red counters, 5 blue, 3 green and 1 yellow. If one counter is drawn, what is the probability that it is:

a yellow   b red   c blue or green

d red, blue or green   e not blue

4 A boy collects marbles. He has the following colours in a bag: 28 red, 14 blue, 25 yellow, 17 green and 6 purple. If he draws one marble from the bag, what is the probability that it is:

a red   b blue   c yellow or blue

d purple   e not purple

5 The probability of a girl drawing a marble of one of the following colours from her bag of marbles is:

blue 0.25   red 0.2   yellow 0.15   green 0.35   white 0.05

If there are 140 green marbles, how many of each other colour does the girl have in her bag?

6 My six tickets for a raffle give me a 0.02 probability of winning first prize. How many tickets have been sold?

7 The probability of getting a bad egg in a batch of 400 is 0.035. How many bad eggs are likely to be in a batch?

8 In a lottery for a new car, 25 000 tickets are sold. My chance of winning is 0.00008. How many tickets do I have?

Remember:
To calculate the probabilities of combined events, multiply the individual probabilities together.

Combined events

When we looked at **combined events** we are looking at problems involving two or more events. The outcomes of these events are usually shown either in *list* form or in a **two-way table**.

Example

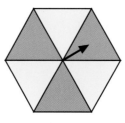 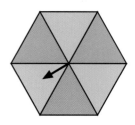

spinner 1   spinner 2

a Two hexagonal spinners are spun. List all the possible outcomes.
b Represent all the possible outcomes in a two-way table.
c What is the probability of getting two reds when the spinners are spun?

a red, red
    red, green
    yellow, red
    yellow, green

b

		spinner 1	
		red	**yellow**
spinner 2	**red**	red, red	red, yellow
	green	green, red	green, yellow

c As the outcomes are all equally likely, P(red, red) $= \frac{1}{4}$.

Exercise 18.3

1 a Two fair coins are flipped. Write a list describing all the possible outcomes.
  b Draw a two-way table to show all the possible outcomes of flipping the two coins.
  c Using your results above, calculate the theoretical probability of getting two Tails.
2 a Two fair tetrahedral dice are thrown. If each is numbered 1–4, write a list of all the possible outcomes when rolling the two dice.
  b Present all the possible outcomes in a two-way table.
  c What is the theoretical probability of both dice showing the same number?
  d What is the theoretical probability that the number on one dice is double that on the other?
  e What is the theoretical probability that the sum of both numbers is a prime number?

3 The letters R, C and A can be combined in several different ways.
  a Write all three letters in as many different combinations as possible.
  If a computer writes these three letters at random, calculate the probability that:
  b the letters will be written in alphabetical order
  c the letter R is written first
  d the letter C is written after the letter A
  e the computer will write the word 'CAR'.
4 **a** Two fair dice are rolled. Draw a two-way table to show all the possible combinations.
  Calculate the theoretical probability of getting:
  b a double 3
  c any double
  d a total score of 11
  e a total score of 7
  f an even number on both dice
  g scores which differ by 3.

Effect of sample size on the result of an experiment

We know that theoretical probability is only a guide as to what we think should happen, not a guarantee of what will happen. However, what happens to actual results of an experiment the more times we repeat it?

Paul and Louise are conducting an experiment. They flip a coin five times and record the number of Heads and Tails they get. They also write their results as a fraction of the total number of times they flipped the coin. They do the experiment three times. The table below shows their results:

	number of Heads	number of Tails	fraction of Heads in total	fraction of Tails in total
first 5 results	1	4	$\frac{1}{5}$	$\frac{4}{5}$
next 5 results	3	2	$\frac{4}{10}$	$\frac{6}{10}$
next 5 results	3	2	$\frac{7}{15}$	$\frac{8}{15}$

Exercise 18.4

1 For this experiment you will need a fair coin.
  a Conduct a similar experiment to the one described above. Repeat it 10 times so that by the end you will have flipped the coin 50 times. Record your results in a table.
  b Can you draw any conclusions from your results?

You will need:
• coin
• two dice

2 For this experiment you will need two fair dice.

a The two-way table below shows all the possible outcomes of adding the scores of the two dice together. Copy and complete the table.

+		dice 1					
		1	**2**	**3**	**4**	**5**	**6**
dice 2	**1**	2					
	2			5			
	3						
	4						
	5						
	6				10		

b What is the probability of getting a total score of 7?

c What is the probability of getting a total score of 12?

d Using the results from the table above, write the theoretical probability of getting each of the total scores shown. Present your answers in a table as shown below.

total score	2	3	4	5	6	7	8	9	10	11	12
fraction		$\frac{2}{36}$							$\frac{3}{36}$		

e Roll two dice 36 times and record the total scores in a frequency table.

f How do the experimental results compare with the theoretical results in the table above?

g Combine all the results from your class and compare the experimental and theoretical results again. Comment on any changes.

h Comment generally on the accuracy of experimental data the more times an experiment is repeated.

SUMMARY

By the time you have completed this chapter you should know:

■ that a **probability scale** ranges from 0 (impossible) to 1 (certain), for example a weather forecast for January might give the chance of snow as 60% or 0.6

■ that if there are n equally likely **outcomes**, the probability of each happening is $\dfrac{1}{n}$, for example the probability of throwing a 5 with a fair dice is $\frac{1}{6}$, as there are six possible outcomes of which only one is a 5

■ how to present **two-way tables** to help calculate the probabilities of two **combined events**, for example a two-way table to show the possible outcomes of spinning two coins is:

		coin 1	
		Head	Tail
coin 2	Head	HH	HT
	Tail	TH	TT

so the theoretical probability of two Heads is $\frac{1}{4}$

■ what is meant by **theoretical probability** and how it may differ from experimental results, for example the theoretical probability of two Heads above is 1 out of 4. But if you only tried the experiment four times, it is unlikely that you would get the results HH, HT, TH, TT

■ that in **experimental probability**, the bigger the sample, the more reliable the results, for example if you continued the experiment above for 400 throws, you would be closer to 100 HH, 100 HT, 100 TH, 100 TT; if you made 40 000 throws, you would get even closer to the theoretical probability.

Exercise 18A

1 What is the probability of throwing the following numbers with a fair dice?
   a a two   **b** an odd number   **c** less than 5   **d** a seven

2 If you have a normal pack of 52 cards, find the probability of drawing:
   a a diamond   **b** a six   **c** a black card   **d** a picture card   **e** a card less than 5

3 150 tickets are sold for a raffle. What is the probability of winning the first prize if the following number of tickets are bought?
   a 1   **b** 5   **c** 20   **d** 75   **e** 150

4 A bag contains 11 blue, 8 red, 6 white, 5 green and 10 yellow counters. If one counter is taken from the bag, find the probability that it is:
   a blue   **b** green   **c** yellow   **d** not red

5 The probability of drawing a red, blue or green marble from a bag containing 320 marbles is:

   red 0.5   blue 0.3   green 0.2

How many marbles of each colour are there?

Exercise 18B

1 An octagonal spinner has the numbers 1–8 on it, as shown below.

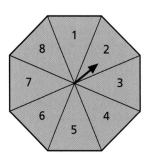

Find the probability of spinning:
a a 7   **b** an even number   **c** a factor of 12   **d** a 9
2 180 tickets are sold for a raffle. What is the probability of winning first prize if the following number of tickets are bought?
a 1   **b** 9   **c** 15   **d** 40   **e** 180
3 An ordinary fair dice and a fair coin are thrown together. Draw a two-way table to show all the possible outcomes.
4 The letters A, B, C and D are written down in a random order. List all the possible combinations of writing the four letters.
5 The probability of drawing a red, blue or green marble from a bag containing 320 marbles is:

red 0.4   blue 0.25   green 0.35

If there are no other colours in the bag, how many marbles of each colour are there?
6 If I buy one ticket for a lottery and have a 0.00002 probability of winning first prize, how many tickets have been sold for the lottery?

Exercise 18C

Investigate, using two-way tables where appropriate, the following problems.

- Two ordinary dice are rolled and the scores added together. What is the most likely total and what is the probability of getting it?
- Two four-sided dice are rolled and the scores added together. What is the most likely total and what is the probability of getting it?
- Two nine-sided dice are rolled and the scores added together. What is the most likely total and what is the probability of getting it?
- Can you generalise the above results for two *n*-sided dice?
- A four-sided and a six-sided dice are rolled and the scores added together. What are the most likely totals and what is the probability of getting them?
- Can you generalise when an *m*-sided and an *n*-sided dice are rolled together?

Exercise 18D

Using card, construct a hexagonal spinner. Colour each of the six sections differently – an example is given here.

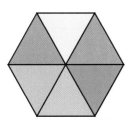

You will need:
- card
- scissors
- colouring pens
- small pencil, or similar pointed stick
- computer with spreadsheet package installed

By putting a short pencil through the middle of the spinner you will then be able to spin it round.

- Spin the spinner 10 times and transfer the results onto a spreadsheet.
- Spin the spinner 100 times and transfer the results onto a spreadsheet.
- Using the graphing facility of the spreadsheet, use graphs of your choice to compare the two sets of results. Write a brief conclusion of your findings.

An example of how you may wish to set out your spreadsheet is given below:

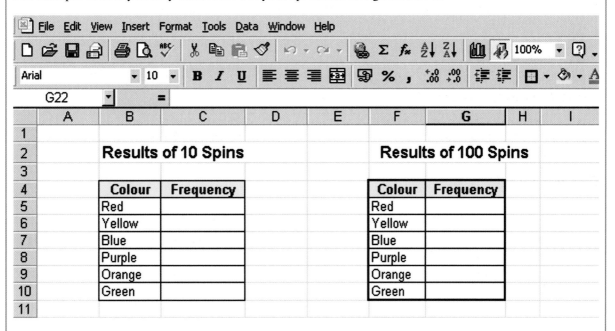

Exercise 18E

Tom Stoppard's play *Rosencrantz and Guildenstern are Dead* sets its tone from a probability experiment in Act 1.

In the book *The Hitchhiker's Guide to the Galaxy*, the spaceship piloted by Zaphod Beeblebrox zipped around the universe with the power of its improbability.

The financier George Soros made £1 billion in a few days by backing the 'balance of probabilities'.

The modern state of Monaco in the South of France was founded on the fact that a little silver rolling ball is as likely to stop in one slot as in another.

You will need:
- computer with internet access or CD-ROM encyclopaedia

Using the internet or a CD-ROM as a resource, find out more about the statements above. Try to answer the following questions.

1 What earlier play does the title of Tom Stoppard's play originally come from?
2 For which academy award winning film did Tom Stoppard write the screenplay?
3 What is the connection between the answers to questions 1 and 2 above?
4 What event did George Soros back and what part did it play in 'Black Wednesday'?
5 Where is the little silver ball rolling?
6 Who wrote the book (which later became a TV series) that included Zaphod, Ford Prefect, Arthur Dent and Marvin the paranoid android?
7 For which medium was the series originally designed?

INDEX

Index